EVOLUTIONARY PERSPECTIVES ON PREGNANCY

EVOLUTIONARY PERSPECTIVES ON PREGNANCY

JOHN C. AVISE

ANIMAL DRAWINGS BY TRUDY NICHOLSON

COLUMBIA
UNIVERSITY PRESS
NEW YORK

Columbia University Press
Publishers Since 1893
New York   Chichester, West Sussex
cup.columbia.edu
Copyright © 2013 Columbia University Press
All rights reserved

The author's research and writing are supported by funds from the University of California
at Irvine.

Library of Congress Cataloging-in-Publication Data
Avise, John C.
  Evolutionary perspectives on pregnancy / John C. Avise with animal drawings by Trudy
Nicholson.
       pages cm
  Includes bibliographical references and index.
  ISBN 978-0-231-16060-5 (cloth : alk. paper) — ISBN 978-0-231-53145-0 (ebook)
  1. Pregnancy in animals. 2. Vertebrates—Reproduction. 3. Invertebrates—Reproduction.
4. Sexual selection in animals. 5. Evolution (Biology) I. Nicholson, Trudy H. II. Title.

  QP251.A95 2013
  573.66—dc23

2012029371

c 10 9 8 7 6 5 4 3 2

References to Internet Web sites (URLs) were accurate at the time of writing. Neither the
author nor Columbia University Press is responsible for URLs that may have expired or
changed since the manuscript was prepared.

To the three women in my life—my mother, Edith; my wife, Joan; and my daughter, Jennifer—who have given me personal but very different exposures to what human pregnancy entails

CONTENTS

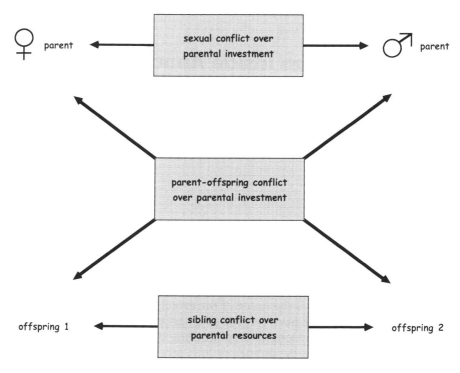

FIGURE 1.1   Three potential sources of intrafamilial conflict in species that display pregnancy (after Parker et al. 2002).

to the phylogenetic histories of species and to the ontogenetic histories (developmental profiles) of individuals, two major biological arenas that are intertwined evolutionarily (Gould 1977). The pregnancy phenomenon prompts many additional evolutionary questions such as this: how are natural selection (arising in this case from interactions between parent and fetus) and sexual selection (stemming from differences in mating success) affected by a syndrome that imposes on one sex much more so than on the other? With a few notable exceptions detailed later, pregnancy is a burden normally borne exclusively by females. However, this huge reproductive obligation is accompanied by a strong evolutionary bias toward feminine control over key biological decisions about species' reproductive modes. Thus, even standard female pregnancy is a double-edged evolutionary sword for both genders, simultaneously conferring upon females much of the reproductive responsibility but thereby in some ways also empowering them while depriving males of much reproductive authority. For these and other reasons, species that display "male pregnancy" are of special interest to evolutionary biologists because in certain cases conventional reproductive roles are reversed.

Thus, when viewed from a comparative perspective, pregnancy speaks not only to many issues relevant to human health but also to the broader ecology and evolution of animal mating systems and a wide range of associated reproductive topics. Finally, pregnancy-like phenomena in nature have diverse natural histories that are fascinating in their own right, both empirically and conceptually. Therein lie the evolutionary themes that this book explores.

The Sexual Life Cycle

The reproductive life cycle of any sexual species is an endless sequence of four signature events (fig. 1.2): (I) gamete production (*gametogenesis*), (II) gamete deployment, (III) gamete union (*syngamy*), and (IV) development (*ontogeny*) of progeny eventually into an adulthood that culminates in aging and death. In each organismal generation, males and females generate haploid sex cells and then deploy these gametes in ways that promote fertilization or syngamy (from "syn," meaning "together," and "gamos," meaning "marriage") to initiate a new generation of offspring. In other words, each fertilization or conception forges a diploid cell (known as a *zygote*) that then proliferates mitotically into a new individual, in which gametogenesis will again take place and restore the life cycle's haploid phase. The hereditary loops in this iterative process stretch unbroken across successive organismal generations like continuous spirals in a Slinky toy or a coiled spring. In some species, pregnancy is an important component of ontogeny (for both parent and child). Conventionally, pregnancy is defined as a period of internal gestation that precedes the birth of free-living progeny. As such, pregnancy constitutes an intriguing biological arena for evolutionary investigations because it is the only time during the sexual life cycle when members of consecutive generations are physically nested inside one another. Furthermore, pregnancy becomes an even more

Factoid: Did you know? Each human body comprises about 50 trillion somatic cells that all trace back to one diploid cell (the zygote) that initiated each pregnancy. About 6 billion people are alive today, so the total number of human somatic cells on earth is an astronomical 300 sextillion (300,000, 000,000,000,000,000,000). Coincidentally, 300 sextillion is also what physicists estimate the total number of stars in the known universe to be.

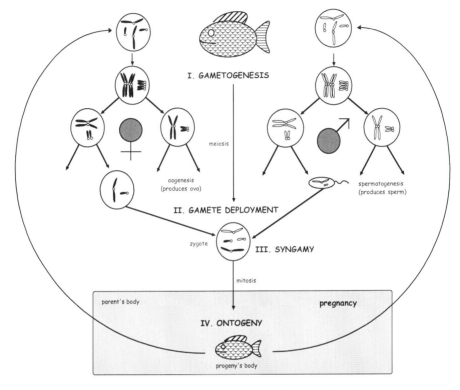

FIGURE 1.2   Signposts in each generational cycle of sexual reproduction. During gametogenesis, meiosis generates haploid sex cells in males and females. Later, during ontogeny, mitotic events proliferate diploid cells in each offspring that was conceived by the union (syngamy) of parental gametes from its sire and dam.

engaging topic for comparative evolution when we expand its definition to include additional categories of embryonic gestation and parental investment in the young.

The four major stages in the sexual life cycle are sequential, so where and when each takes place in a given species influences how the successive stages play out. The following sections hint at the diverse locations of gametic production, deployment, union, and embryonic development across a potpourri of animal and plant species.

Gametogenesis

This process is initiated by diploid cells (primary spermatocytes and oocytes) in the testicular and ovarian tissues of animals or by analogous cells in the

> **Factoid: Did you know?** Each human zygote is a single cell about the size of the period at the end of this sentence.

male and female flowers of plants. The cells that enter meiosis trace their own mitotic ancestries back to the zygote, which was also the ultimate progenitor of all other cells in each organism's body or soma.

When a multicellular organism begins to grow by mitotic cell divisions from a fertilized egg, there comes a time when its germ line (which is destined to produce gametes) segregates from the soma (diploid cell lines that otherwise make up the body). In many animals, this sequestration occurs quite early in development and leads to the gonadal tissues of males and females, wherein spermatogenesis and oogenesis, respectively, transpire. In plants, the sequestration between germ line and soma may occur later, so, until meiosis begins, there may be no evident distinction between an individual's somatic and germ-line cells. Thus, with regard to cellular origins and bodily locations of reproductive tissues, plants generally have more developmental flexibility (*phenotypic plasticity*) than do most animals. In other words, with regard to reproduction, plants have relatively broad "norms of reaction" during ontogeny. Nevertheless, even vertebrate animals show considerable variation from group to group with respect to when and where gametogenesis occurs. For example, in bony fishes the testes and ovaries derive during ontogeny from a single precursor tissue that can differentiate rather flexibly during an individual's lifetime, thereby helping to account for the great diversity of sex-determination mechanisms in fishes (DeWoody, Hale, et al. 2010; Mank et al. 2006; Mank and Avise 2006b, 2009) compared to the uniformity of sex-determination modes in mammals or birds (box 1.1).

> BOX 1.1   Sex-Determining Mechanisms
>
> Most mammals, including humans, have a chromosomal mode of sex determination, in which each "heterogametic" male carries two types of sex chromosomes (X and Y) and each "homogametic" female carries two copies of the X. Thus, during gametogenesis each haploid sperm cell receives either an X chromosome or a Y chromosome with approximately equal likelihood, whereas every ovum receives an X. Whether a boy or a girl is conceived during syngamy then depends on whether an X-bearing or a
>
> *(continued)*

BOX 1.1   (*continued*)

Y-bearing sperm cell fertilizes the oocyte. Sex determination in birds is likewise chromosomal, but the rules are reversed: females are the heterogametic sex (conventionally designated ZW), males are homogametic (designated ZZ), and each offspring's sex registers whether the relevant oocyte from its mother was Z bearing or W bearing.

Mechanisms of sex determination are collectively far more varied in other vertebrate and invertebrate taxa. In fishes, for example, various species are known to display male heterogamety (much like mammals), female heterogamety (much like birds), combinations of XY-like and ZW-like systems (Cnaani et al. 2007; Vicari et al. 2008), or various expressions of monogenic or polygenic rather than strictly chromosomal sex determination (Mair et al. 1991; Vandeputte et al. 2007). Indeed, some piscine species have dispensed almost entirely with direct genetic control over gender, relying instead at least partly on environmental cues. In some fish species, for example, the temperature of incubation strongly influences the sex of the progeny (Mair et al. 1980; Desperz and Melard 1998; Baras et al. 2001). The general fluidity of sexual ontogeny in fish is also illustrated by hermaphroditic species, in which an individual functions simultaneously as male and female or perhaps switches back and forth between male and female during its lifetime, often depending on the social environment (Avise 2011).

In essentially all sexual species, haploid reproductive cells that emerge from meiotic events come in two distinct size classes that define the two genders. The small male gametes of plants and animals are called pollen (technically, male gametophytes) and sperm, respectively, whereas the much larger female gametes in all species are known as ova, oocytes, or unfertilized eggs. This size difference, termed *anisogamy*, arose hundreds of millions of years ago in the early evolutionary history of multicellularity and sexual reproduction (Majerus 2003), apparently via disruptive selection pressures that favored the union of large and relatively immobile sex cells with smaller and more mobile ones (box 1.2). Today, strongly bimodal distributions of gamete size and mobility continue to characterize extant multicellular species that reproduce sexually. Indeed, gamete size is the only phenotypic feature that consistently distinguishes males from females in any plant or animal taxon. By definition, individuals that produce relatively small gametes are males, whereas those that produce the bulkier gametes are females. The same distinction holds true even for dual-sex individuals in hermaphroditic species. By definition, a hermaphrodite functions at any specified time as a male or as a female depending, respectively, on whether the gametes that it deploys at that time are either small and mobile or large and immobile (Avise 2011).

BOX 1.2   Evolutionary Origins of Anisogamy and Separate Sexes

Anisogamy is the pronounced difference in size (and often in mobility) between male and female gametes. This ancient condition probably originated around the same evolutionary time as did multicellularity and sexual reproduction. How this bimodal distribution of gametes (large versus small) first came about has been the subject of much informed theorizing.

In an evolutionary scenario developed by Parker et al. (1972), a primeval "competition" among gametes precipitated the original transition from isogamy (the presumed ancestral condition of equal-sized gametes) to anisogamy. Imagine an ancestral population of isogamous organisms in which a new mutation initially led one individual to produce smaller than normal gametes. Because its sex cells were smaller, that organism could produce more gametes with the same energetic investment, so it would have enjoyed an initial fertilization advantage over its compatriots who still produced larger gametes. The small-gamete mutation would thus increase in frequency in the population. As the mutant allele became more common, the likelihood increased that two small gametes would meet and fuse. The resulting zygote would likely be debilitated, however, because the small gametes that produced it offered few nutrients for the nascent embryo. Thus, natural selection would favor any tendency for small gametes to fertilize only large ones, and the resulting competition among small-gamete individuals for successful fertilization would select for individuals who produced even smaller (and thus more) gametes. As tiny gametes became more prevalent, selection pressures escalated on large-gamete individuals to produce larger gametes to compensate for the limited nutrients that the tiny gametes contributed to a zygote. From this disruptive-selection regime favoring both large and small gametes, anisogamy eventually emerged as an evolutionarily stable outcome.

Hurst (1990) offered a different scenario for the evolutionary origin of anisogamy. It is well known that intracellular parasites such as bacteria often inhabit the cellular cytoplasm and that they can impose strong selective

(continued)

BOX 1.2   (*continued*)

pressures on hosts (Burt and Trivers 2006). Hurst argued that anisogamy evolved because of an advantage it conferred (relative to isogamy) in reducing the probability of infection by a disease agent during fertilization. Today, small male gametes contribute little cytoplasm to the fertilized egg, which instead acquires cytoplasm mostly from the ovum. Because only the female contributes appreciable cytoplasm to a zygote, anisogamy diminishes the probability that a zygote acquires a cytoplasmic agent of a sexually transmitted disease via syngamy. This advantage probably also applied early in evolutionary history and perhaps contributed to selection pressures favoring anisogamy.

A third hypothesis for the evolution of anisogamy also builds on the observed disparity between male and female gametes in cytoplasmic contributions to the zygote. Mitochondria are organelles that reside by the hundreds or thousands in the cytoplasm of each somatic and germ cell. They carry tiny genomes that encode key components of the molecular machinery by which cells produce chemical energy. Mitochondria are not cellular parasites, but, like cytoplasmic bacteria, they normally are transmitted via the oocyte from one organismal generation to the next. This uniparental mode of maternal inheritance led Hurst and Hamilton (1992) to propose that anisogamy was (and still is) favored by natural selection because it minimizes the potential for intrazygote conflict between what would otherwise be genetically distinct populations of cytoplasmic organelles delivered by the fusing gametes. Similar kinds of evolutionary arguments in favor of anisogamy can be made with respect to a plant's cytoplasmic organelles, which in addition to mitochondria include chloroplasts, which also carry their own little genomes.

These three hypotheses for the origin of anisogamy are not mutually exclusive, so all of them probably contributed to the evolutionary emergence and maintenance of the phenomenon.

Factoid: Did you know? Although biologists speak of female sex cells as being the larger class of gametes (compared to those of males), an oocyte is tiny in absolute terms, barely visible to the naked human eye.

Gamete Deployment

After producing sex cells, each prospective parent must deploy them in ways that promote their union with gametes from one or more members of the opposite sex. Males and females often approach this task in very different ways, however, due ultimately to differences in gamete size and mobility that accompany anisogamy. Sperm cells typically have tails for swimming, and male gametophytes in most plants (while not necessarily motile) are highly mobile when dispersed as pollen. By contrast, the ova of plants and most animals (with some notable exceptions) tend to be sedentary, typically remaining cloistered inside the ovules of female flowers or nestled in the innermost reproductive tracts of female animals. Thus, whereas females usually retain their ova internally at least for some time, males almost invariably expel gametes from their bodies. By safeguarding her precious ova until a proper opportunity for fertilization arises, each female in effect is wisely managing her valuable genetic assets; moreover, by dispensing his inexpensive gametes with relative abandon (i.e., by "sowing his oats"), each male also is striving to improve his odds in the reproductive sweepstakes. This basic dichotomy of reproductive tactics (gamete shedding by males versus gamete retention by females) is a major reason that the female sex is predisposed to evolve pregnancy-like phenomena. Internal gestation normally starts with stay-at-home female gametes that await fertilization by their more adventuresome male counterparts.

On the other hand, in certain species the females, too, seem cavalier about dispersing their gametes. I am referring now to the many marine invertebrates and pelagic fishes in which females and males broadcast large numbers of relatively naked *gametocytes* (those without heavy protective casings) into the sea, trusting mostly to chance that some will encounter gametes from the opposite sex. To improve the odds of fertilization, adults typically release species-specific pheromones (out-of-body chemical signals) or employ environmental cues (such as photoperiods or tidal cycles) to synchronize reproductive efforts. In many such "broadcast spawners," a further coordination of sexual efforts can take place when individuals of both sexes stockpile large numbers of gametes in special body cavities (often the coelum) before releasing them synchronously in mass spawning. Although broadcast spawning by both sexes is common in marine invertebrate species, it is virtually nonexistent in freshwater and terrestrial invertebrates, presumably because osmotic stresses and desiccation compromise gametic survival in these two respective environments.

Broadcast spawning by females normally entails external fertilization and no parental care of the young (i.e., no pregnancy), so henceforth in this book we can mostly disregard this mode of gametic deployment. Not so easily dismissed, however, are the many animal and plant species in which male gametes alone are spewed into the environment either by males or by her-maphroditic individuals. In wind-pollinated plants and in "spermcasting" (Bishop and Pemberton 2006) marine invertebrates, such individuals release pollen or sperm that then emigrate through air or water to gravid but sessile females. If and when fertilization then takes place within the female body, opportunities arise for prolonged embryonic gestation by dams. For marine invertebrates, the term *brooding* (rather than pregnancy) is often employed, but the net effect is similar, especially when females offer additional resources to their brooded young (chapter 4).

In most other animal species, males offer much more personalized repro-ductive services to females. One such sexual service is direct "home-delivery" of sperm via copulation and ejaculation, the standard prelude to internal fertilization in most viviparous (live-bearing) vertebrates (chapter 2) and other animals with internal female pregnancy. Another personalized but less intimate approach entails partner spawning without intromission or insemi-nation, a standard prelude to external fertilization in many oviparous (egg-laying) fishes, amphibians, and other animals (chapter 3). A third approach—prepackaged sperm drop-off—is an even less personalized method of gametic transfer. In many invertebrates and some vertebrate species, a male exter-nally deposits a sperm packet (*spermatophore*), which a female then picks up from the substrate (perhaps using her cloacal lips, as in many primitive sala-manders with internal fertilization; Wake 1993). What a female does next with such packaged gifts varies tremendously among species. In some cases she might use the sperm merely as nutritional supplements, like vitamin pills, either to digest herself or perhaps to feed to her gestating offspring (Dobbs 1975; Gardiner 1978); in other cases she might store viable sperm (in special-ized compartments known as *spermathecae*) for as long as several years before using them to fertilize her ova.

In the botanical world, male and female plants do not copulate directly, but many of them do so vicariously by employing animal intermediaries (such as insects, birds, or bats) that they attract and bribe with sensual flowers and nu-tritious nectars and pollen. By delivering pollen from one flower to another, pollinators are recompensed mediators of plant "mating events." Thus, just as surely as do animals that copulate, animal-pollinated plants have mating sys-tems (and body parts, including flowers) that are subject to both sexual and natural selection. Indeed, the mating games played by such plants can be es-pecially Machiavellian because complicated selection pressures often underlie the coevolutionary dances between plants and their animal pollinators. In any event, next time you go outdoors, try reflecting on the following truth: a great majority of nature's intriguing sights, sounds, and odors (including many more that are beyond human senses) are proximate signals by which males and females of various plant and animal species communicate with one another with the ultimate end of uniting conspecific gametes.

Overall, the diversity of ways in which sexual animals and plants deploy their gametes can seem bewildering, but all such routes conform to one universal anisogamy-related truth: wherever members of the female sex present their precious ova for fertilization, conspecific members of the male sex (or at least their gametes) are sure to follow.

Syngamy

The next crucial step in each sexual life cycle is the union of male and female gametes during fertilization. As mentioned earlier, syngamy normally takes place at a location that is either removed from both parents (as in broadcast spawners) or within or near the body of the dam (in the many species in which ova are retained by the female sex and some form of female pregnancy often ensues). Much less common is the converse situation, in which females deliver and entrust their gametes to particular males. Nevertheless, this is precisely what happens when a female pipefish or seahorse (family Syngnathidae) deposits her unfertilized eggs into the brood pouch of a male, who then fertilizes the ova internally and incubates the embryos for several weeks until giving live birth (chapter 7). The syngnathids provide otherwise rare examples of syngamy within a male's body and they illustrate the phenomenon of male pregnancy in quintessential form. By analogy, the many nest-tending fish species can be deemed to display "external male pregnancy" (Avise and Liu 2010). In such cases a female lays her oocytes into the nest of a "bourgeois" (shopkeeper) male who then fertilizes the eggs and tends to the embryos externally (chapter 7). Finally, "external female pregnancy" might be said to occur when a female lays oocytes or fertilized eggs and then incubates progeny outside her body, a common practice in many invertebrate animals.

In any species, syngamy marks the beginning of the next generation and introduces a new genetic participant—the progeny—to each unfolding reproductive drama. Each fertilization is especially consequential in species that display pregnancy, in part because an offspring who then gestates or otherwise receives care from its dam or sire inevitably creates a family triangle with inherent genetic conflicts of interest (as well as new mutualities of interest) among the three genetically involved parties (dam, sire, and progeny). Part II of this book explains how this social complexity plays out during various forms of pregnancy.

Ontogeny and Pregnancy

Every zygote requires nourishment as it begins to divide and multiply into a multicellular embryo. This sustenance comes initially from organic materials (yolk) deposited in the ovum's cytoplasm (box 1.3), usually supplemented with other foodstuffs (notably an endosperm in flowering plants) that the dam has provisioned for her progeny, and the source of these comestibles then influences how ontogeny proceeds in each taxon. Yolk deposition (*vitellogenesis*) is the maternal deposition of nutritional resources into an egg, a process that occurs in most animal species (with the notable exception of placental mammals). Each embryo then requires additional resources as its development continues, with the magnitude and source of the supplementation helping to define the type of pregnancy (or lack thereof) of a given species.

BOX 1.3   The Cellular Cytoplasm

All multicellular algae, plants, fungi, and animals are composed of eukaryotic cells, in which a semipermeable membrane demarcates the nucleus from the cytoplasm. The nuclear compartment houses chromosomes that carry each organism's nuclear genome, whereas the cytoplasm houses nearly everything else, including mitochondria (biochemical "factories" that generate energy in all eukaryotic cells) and chloroplasts (key enablers of photosynthesis in plants and algae). These two organelles carry their own little snippets of genetic material, known as mitochondrial (mt) DNA and chloroplast (cp) DNA, respectively, which are closed-circular molecules, each a few tens of thousands of nucleotides long. Although these two cytoplasmic genomes are individually tiny in comparison to the nuclear genome (which typically has several billion base pairs), they occur in great abundance in some cells. For example, the cytoplasm of a mature oocyte may house several hundred thousand mtDNA molecules, whereas a sperm cell typical carries only about 100 such copies. The cellular cytoplasm is home to many other biochemical substances and processes as well, including legions of organelles known as *ribosomes*, which translate nucleic acids into proteins that conduct each organism's daily metabolic chores.

When a male gamete and a female gamete unite during fertilization, the vast majority of the zygote's cytoplasm comes from the ovum. This basic sexual asymmetry, due to anisogamy, helps explain why mtDNA and cpDNA are maternally inherited in most animal and plant species. It also helps to account for the extraordinary control that the female sex has over "periconception" (Bourc'his and Voinnet 2010), which encompasses gametogenesis, syngamy, and early zygotic development. During this crucial time in the sexual life cycle, "cytoplasmic determinants" (proteins, RNA molecules, and other substances that the mother produced and deposited in the ova) play key roles in regulating gene expression, which affects the early development of the embryo.

For species with internal fertilization and embryonic development inside a pregnant dam or sire, a parent's investment in its offspring can soar during the gestational phase of the life cycle. In viviparous mammals and also in many live-bearing fishes, a placenta delivers massive amounts of nutrition to the fetus, as evidenced by two facts: (a) embryos in such taxa typically gain weight during a pregnancy, and (b) each newborn must make new feeding arrangements (such as suckling) soon after parturition to compensate for the loss of parental support via the placenta. By contrast, in oviparous (egg-laying) species such as birds and many reptiles, each prehatch embryo must ration the finite yolk and other nutriments that came prepackaged within its harder-shelled egg.

Ontogenetic and Evolutionary Gradations of Pregnancy

Gestational phenomena can be matters of degree (fig. 1.3) that defy attempts to define pregnancy by any single universal criterion (Wourms 1981; Black-

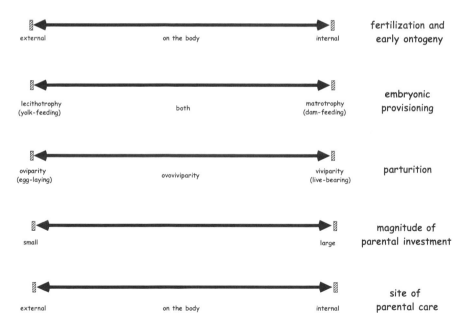

FIGURE 1.3   Several biological dimensions along which different taxa collectively show gradations between nonpregnancy and "pregnancy." Along each dimension, species toward the right side of the continuum conform more closely to traditional concepts of pregnancy. Note that the positions of species along several of these gestational scales often covary.

burn 1994, 2000). However, it is much more important to appreciate the diversity of nature's incubational modes than it is to quibble about formal definitions of pregnancy.

Location of Fertilization

Some biologists might wish to equate pregnancy with internal fertilization under the rationale that in all such cases a parent (usually the dam) internally carries—at least temporarily—one or more cells of the following generation. However, by this definition, all birds and other such oviparous species would have to be counted among the organisms that display pregnancy (which many people might argue is going too far). Any such definition of pregnancy would also exclude all species with external fertilization even if the embryos receive substantial parental care (as in many nest-tending fishes) and even if the embryonic gestation occurs within a parent's body (as occurs, for example, in gastric-brooding frogs and mouth-brooding fishes, described later).

Nature of Parturition

Another argument is that internal fertilization is just a standard prelude to viviparity, which itself should be the criterion for establishing whether a species displays pregnancy. However, viviparity and its traditional antithesis (oviparity or egg laying) are bridged in many taxonomic groups by ovoviviparity (chapter 3), a gestational mode in which an embryo begins development in a shelled egg but then hatches inside its mother before being born alive. Furthermore, among ovoviviparous species the relative duration of an embryo's intradam development while encased in an egg rather than being hatched varies greatly—from nil to almost full term (as defined by parturition in the dam).

Embryonic Provisioning and Parental Care

Another conventional distinction relevant to pregnancy concerns whether an internally gestating embryo receives nutrients solely from yolk (*lecithotrophy*) or whether the dam adds supplements (*matrotrophy*). However, this dichotomy, too, can be strained. For example, embryos in ovoviviparous species typically are succored by yolk initially but later in the pregnancy may tap additional maternal resources more directly. Indeed, these two phases of intraparent gestation are themselves matters of degree. Some ovoviviparous species show delayed hatching and mostly lecithotrophic development; others show early hatching and high matrotrophy; and still others display various other combinations in the source, amount, and duration of nutritional investment by the incubating parent. Thus, conventional oviparity, viviparity, and ovoviviparity form a gradient of gestational modes that make it hard to demarcate species that do or do not display pregnancy. Furthermore, the parental-care gradient can extend well beyond parturition in the many species in which one or both parents provide varying degrees of care to their postpartum embryos.

Overall, given that pregnancy can be a matter of degree in several evolutionary respects, this book adopts a heuristic approach by comparing a wide variety of gestational phenomena.

Viviparous Pregnancy in the Reproductive Timeline

Another way to think about the temporal schema of sexual reproduction in viviparous species is to appreciate the fact that several landmark events— gametogenesis, copulation, conception, implantation, and birth—also demarcate distinct biological arenas where different categories of selection may take place (fig. 1.4). For example, each viviparous pregnancy is flanked by conception and birth and thus constitutes a temporal window in which viability-based natural selection is likely to materialize due in part to the delicacy of genetic and physiological interactions between parent and fetus. By contrast, sexual selection is likely to register in a prezygotic domain

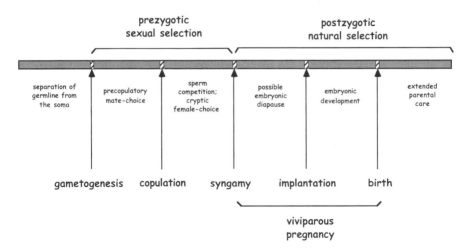

FIGURE 1.4   The position of standard viviparous pregnancy in the temporal schema of organismal reproduction.

bounded by gametogenesis and syngamy. This area itself is split by the cop-ulation event into two major playing fields for sexual selection: precopula-tory mating behaviors and postcopulatory gametic phenomena (discussed later). Various prezygotic phenomena are themselves linked to some extent, such as when multiple mating precedes effective sperm competition within a female's reproductive tract. Pre- and postconception phenomena also have various evolutionary connections, for example, when pregnancy itself gen-erates selection pressures on the mating behaviors of males and females (chapters 7 and 8). Thus, although viviparous pregnancy occupies only a modest fraction of the reproductive timeline in species with internal gesta-tion, it can have major ramifications for how selection processes operate throughout the procreative process.

Heterochrony

Like tabs on a typewriter, the hatched bars in figure 1.4 can shift to the left or right during the evolutionary process in response to selection pressures that widen or narrow the ontogenetic arenas that these tabs demarcate. Such shifts in the timing and mode of ontogeny are called "heterochrony," a gen-eral evolutionary phenomenon that affects numerous biological processes, including various pregnancy-related events. For example, when the tab for birth shifts to the left such that parturition falls earlier rather than later in ontogeny, embryonic development in a genetic lineage might undergo an evolutionary transition from standard viviparity (with mostly internal em-bryonic gestation) to oviparity (with mostly external embryonic develop-ment). Heterochronous changes of this sort underlie evolutionary shifts along a gestational continuum that ranges from viviparous live birthing (chapter 2) to ovoviviparity to oviparous egg laying (chapter 3).

Heterochronous shifts in ontogeny during evolution are often promoted by natural selection (chapter 6) and sexual selection (chapters 7 and 8), but causality also occurs in the reverse direction: heterochronous shifts in the ontogenetic timeline affect the operation of different forms of selection. For example, a common shift of ontogenetic tabs occurs when copulation and syngamy (fig. 1.4) move farther apart in time, as routinely happens when a mated female stores sperm before using them to fertilize her eggs (box 1.4). Female sperm storage can widen the temporal window for prezygotic sexual selection via postcopulatory sperm competition and/or cryptic female choice (box 1.5). Similarly, any shift of the ontogenetic tabs that widens the gap be-tween gametogenesis and copulation (fig. 1.4) might enhance the opportu-nity for precopulatory mate choice, another standard form of prezygotic sex-ual selection (chapter 7).

Sperm Storage by Females

The reproductive tracts of many female animals include special internal organs that can store viable sperm for considerable periods of time. For example, a fruit fly can utilize stored sperm to fertilize her eggs for up to two weeks following copulation. The same is true for many bird species. In most mammals, females can store and use viable sperm for only a few days at most, but some bats are able to do so for as long as 30 weeks postmating, and some female lizards can store sperm for two months or more before fertilizing their eggs. Likewise, several months may intervene between copulation and egg laying in some newts and salamanders. In some fish species that have internal pregnancies and bear live young, females can stockpile viable sperm for 3–10 months after mating. Even more impressive, however, are animals whose feats of gamete hoarding are measured in years. These include some social insects in which a queen bee or ant continues throughout her life to produce offspring, using sperm stored since her wedding flight.

Among vertebrate animals, the endurance champions of sperm storage are female turtles and snakes, who can employ sperm from bygone nuptial affairs for 5 years or more. Such long-term sperm storage came to light from the observation that captive females housed in isolation sometimes continued to produce offspring long after they had last been with a male. Unless these progeny arose via virgin birth (*parthenogenesis*), these females must have used sperm stored from a carnal liaison that had taken place years before. In recent years, geneticists have confirmed that successful long-term female sperm storage in reptiles can indeed happen both in captivity and in the wild (see Avise et al. 2002).

Sperm Competition and Cryptic Female Choice

In his second most famous book, *The Descent of Man, and Selection in Relation to Sex*, Darwin (1871) defined sexual selection as the "advantage which certain individuals have over other individuals of the same sex and species, in exclusive relation to reproduction." Darwin stated that sexual selection could be mediated both by intrasexual competition or combat (especially among males) and by intersexual preferences (e.g., female choice

(continued)

BOX 1.5 *(continued)*

of attractive mates). He also noted that sexual selection could function in opposition to natural selection with respect to particular phenotypic traits (such as a peacock's tail), but he generally viewed sexual selection as being less effective than natural selection.

Today, we far better appreciate not only that sexual selection's two cornerstones—"male-male competition" and "female choice"— can be extremely powerful evolutionary shaping forces but also that these two processes routinely operate both before and after copulation, a fact that uncharacteristically escaped Darwin's notice (Eberhard 2009). When these two modes of sexual selection take place inside the female reproductive tract during the interval between copulation and syngamy, they are termed *sperm competition* and *cryptic female choice* (Baker and Bellis 1995; Birkhead and Møller 1992, 1998; Parker 1970; Smith 1984), two miniature worlds of gametic action that are every bit as fascinating as the macroscopic worlds of mate competition and mate choice, which were the traditional foci of research on sexual selection.

Sperm competition can occur whenever sperm from two or more males directly vie for the fertilization of ova. For species with internal fertilization, sperm competition implies that a female either mated with or otherwise acquired sperm from multiple males (another route being spermatophore take-up). Especially for species with direct intromissive copulation, many morphological characteristics and reproductive behaviors of males have been interpreted as adaptations to the genetic challenges of competing with other males' sperm. For example, in many worms, insects, spiders, snakes, and mammals, a male secretes a plug that serves as a temporary chastity belt to block a female's reproductive tract from subsequent inseminations. In many damselflies and dragonflies (order Odonata), males have a recurved penis, which, when inserted into a female, physically scoops out sperm from males with whom she previously mated. Other widespread male behaviors often interpreted as offering paternity assurance in the face of potential sperm competition include prolonged copulation (up to one week in some butterflies), multiple copulations with the same female, and postcopulatory mate guarding.

From a female's perspective, mechanisms to prevent sperm competition are not necessarily desirable, and this can lead to reproductive conflicts of interest between the sexes (Eberhard 1998; Knowlton and Greenwell 1984), such as when a female seeks the best possible sperm to fertilize her ova even if they did not come from her caregiving partner.

(continued)

BOX 1.5   (*continued*)

Thus, growing evidence suggests that female reproductive tracts often play far more active roles than previously supposed in postcopulatory cryptic female choice of sperm for the fertilizations (Birkhead and Møller 1993).

Some Unorthodox Routes to Internal Pregnancy

Just as the pattern of gamete deployment in each species dictates where fertilizations or conceptions take place, so does the site of syngamy influence the nature and course of any pregnancy-like phenomena that might ensue. Copulation and internal fertilization are standard prerequisites for internal gestation (standard pregnancy) and eventual birth, but even this seemingly obligatory biological connection has a few exceptions, as the following extraordinary examples illustrate.

Internal Gestation Following External Fertilization

In the South American frog *Rhinoderma darwinii* (fig. 1.5) a male amplexes a female by holding onto her back and releasing sperm as she lays ova. Later, he scoops up the externally fertilized eggs in his mouth and deposits them in his vocal sac, where the embryos develop and metamorphose into froglets over a period of several weeks (Busse 1970). The father then spits out his brood. Similar kinds of incubational arrangements characterize many fish species in which a sire or dam carries externally fertilized eggs and embryos in its oral cavity or gill chambers. "Mouth brooding" (chapter 7) has been documented in members of six taxonomic families: Ariidae (sea catfish), Cyclopteridae (lumpfish), Apogonidae (cardinal fish), Cichlidae (Kobmuller et al. 2004), Opistognathidae (jawfish), and Osphronemidae (gouramies) (Helfman et al. 1997). In various cichlids, for example, a sire or dam (or both in some species) collects and holds fertilized eggs in its mouth until the embryos hatch, after which the parent may continue to mouth-carry its fry, especially when danger threatens.

Although some people may disagree that mouth brooding in frogs or fish qualifies as internal incubation (much less as pregnancy), a more dramatic example involving another amphibian is less ambiguous in that regard. In the Australian gastric-brooding frog, *Rheobatrachus silus* (fig. 1.6), a female ingests her externally fertilized clutch of about 30 eggs before temporarily shutting down her digestive system (Tyler et al. 1983) and

FIGURE 1.5   *Rhinoderma darwinii* (Rhinodermatidae), a South American frog in which males brood embryos in their vocal sac.

Factoid: Did you know? The gastric-brooding frog was thought to be common when it was discovered in 1973, but, sadly, within a decade the species seems to have disappeared and may now be extinct.

brooding the young in her stomach. At the end of this 8-week process, the pregnant dam gives "oral birth" by regurgitating metamorphosed froglets, who then simply hop away (Corben et al. 1974; Tyler and Carter 1981).

FIGURE 1.6 *Rheobatrachus silus* (Myobatrachidae), an Australian frog with "gastric pregnancy," in which a female ingests fertilized eggs and gestates the embryos in her stomach.

Virgin Births: Internal Gestation Without Sex

In another nonstandard route to internal pregnancy, some species have dispensed with meaningful sex and syngamy altogether. In about 100 extant vertebrate species (0.1% of the world's total), as well as in some invertebrate animals, embryos are "immaculately conceived" within virgin mothers who have engaged in some form of parthenogenesis (from "parthenos," meaning virgin, and "genesis," meaning origin). Parthenogenesis is a form of clonal reproduction (Avise 2008), so the progeny in these all-female species are genetically identical to their respective dams and typically have neither a father nor paternally derived genes. In such parthenogenetic organisms, females produce nonfertilized but nevertheless diploid eggs that give direct rise to daughters via mitotic cell divisions. Although many parthenogenetic females lay eggs that hatch externally, dams in some other parthenogenetic taxa give live birth to their daughters and therefore surely qualify as having been pregnant. Examples include some sharks (Chapman et al. 2007), a few snakes (Booth et al. 2010), various fish species in the family Poeciliidae (Avise et al. 1992), and many aphid insects.

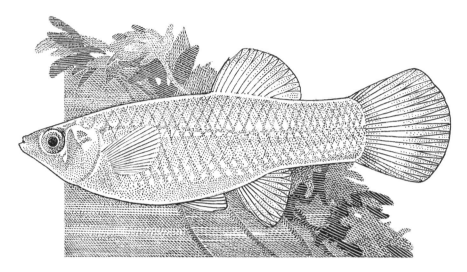

FIGURE 1.7   The Amazon molly, *Poecilia formosa* (Poeciliidae), a viviparous all-female fish species in which pregnant females give "virgin birth" to offspring that are genetically identical to their dams.

Factoid: Did you know? In some parthenogenetic aphids, a pregnant female may carry a daughter who herself quickly produces a daughter, such that three generations "telescope" inside one another like the nested figurines of a Russian doll.

With regard to the unisexual live-bearing fishes, consider, for example, the viviparous Amazon molly, *Poecilia formosa* (fig. 1.7). In this all-female species from streams in northeastern Mexico, each pregnant dam employs a parthenogenesis-like reproductive mode known as *gynogenesis* to produce and then give live birth to progeny that are clonally identical to one another and to their mother. Like all other unisexual vertebrate taxa (Avise 2008), the Amazon molly arose following interspecific hybridization between normal bisexual species, in this case *Poecilia latipinna* and *P. mexicana* (Avise et al. 1991; Stöck et al. 2010). Ever since that date of origin, perhaps 300,000 years ago (Schartl et al. 1995), its clonal lineages have persisted without material genetic input from males.

In rare instances, other unisexual fish species have arisen from more complex hybridization involving more than two ancestral species (Vrijen-

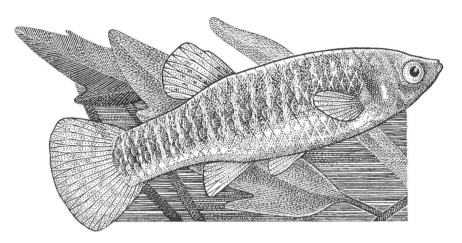

FIGURE 1.8   *Poeciliopsis monacha-lucida-viriosa*, a viviparous unisexual fish species that arose via hybridizations involving three bisexual progenitor species and now propagates by gynogenesis.

Factoid: Did you know? The Amazon molly derives its common name from warrior Amazons, the fictitious tribe of all-female humans in classical mythology.

Factoid: Did you know? Pregnant females in many unisexual taxa are not always strictly virgins. In many such live-bearing species, a female does mate (with males from related sexual species), but the sperm cells she acquires merely stimulate parthenogenetic development of clonal offspring from her technically unfertilized eggs.

hoek and Schultz 1974). One example is the live-bearing *Poeciliopsis monacha-lucida-viriosa* (fig. 1.8) from northwestern Mexico. In this case, the clonal offspring to which the viviparous females give birth carry genes from three ancestral taxa.

Sexual Inequalities and Female Authority Over Reproduction

It would be hard to overstate the evolutionary significance of anisogamy, which directly or indirectly underlies many of the gender-based asymmetries that pervade sexual reproduction. As described earlier, anisogamy promotes gender-specific proclivities (such as gamete retention by females and gamete dispersion by males) that often promote within-female syngamy; which in turn predisposes the female sex to evolve pregnancy-like phenomena; which in turn makes females even more limiting as a reproductive resource compared to males; which in turn amplifies the evolutionary control of reproduction by the female gender, influences the operation of sexual selection, and exacerbates the proverbial "battle between the sexes" (chapter 7).

An ovum's large initial size (compared to that of a sperm cell) comes mostly from its large cytoplasm (see box 1.3) that houses nutrients, as well as many of the chemical instructions necessary to support and direct early mitotic cell divisions of the zygote. A male gamete carries much less cytoplasm and therefore contributes little more than its nuclear genes to a fertilized egg. In many animal species, an ovum's bulk is further enhanced during vitellogenesis when the dam adds yolk substances that make each egg an even richer pantry for her developing embryos. In flowering plants, a functional equivalent of yolk is a starch- or lipid-rich endosperm that nourishes each seed-encased embryo.

Anisogamy implies that the small gametes of males are individually much cheaper to produce than ova, meaning that a male has the inherent capacity to produce vastly more sperm or pollen than a female can produce fertilizable eggs. This potential is realized routinely. For example, a human male during his lifetime spews out more than 10 trillion spermatozoa (500 million per ejaculate), whereas a woman releases into her upper reproductive tract only about 400 ova (on a monthly cycle). Huge numerical inequalities of this general magnitude characterize most sexual species, and they have profound evolutionary ramifications for how males and females conduct their reproductive affairs in their never-ending efforts to enhance personal genetic fitness via successful offspring production.

Because of its sex-specific effects on sexual selection and natural selection, anisogamy has many evolutionary consequences. As elaborated in later chapters, anisogamy almost automatically tips the scales toward stronger sexual selection on males than on females in many species because females and their ova become a limiting reproductive resource over which males and their gametes actively compete for fertilization. Having been consigned by evolution to produce the costlier class of sex cells, females right from the outset have invested more in each zygote than their part-ners have. Over time (during both ontogeny and evolution), this basic asym-metry of investment in turn predisposes females to commit even more time and resources to rearing progeny. Placental pregnancy in mammals is merely one illustration of how this gender-specific feedback process has amplified over evolutionary time into extreme manifestations of maternal devotion and apparent female self-sacrifice on behalf of internally gestated young.

From this evolutionary perspective, anisogamy might be deemed the pri-mal source of female enslavement to procreation (i.e., as the original first step down a long and slippery evolutionary slope that ever since then has

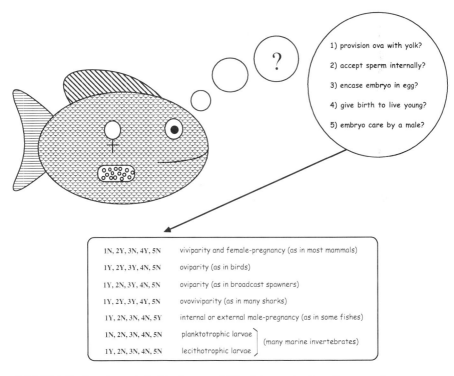

FIGURE 1.9   Female control of (and responsibility for) reproduction. Shown are five evolutionary questions that nature poses mostly to females, all pertaining to how females might handle the gametes and embryos that anisogamy has entrusted to them (see text). The tactics that females have adopted in various phylogenetic lineages help to define alternative modes of reproduction and embryonic development because males must basically accommodate whatever answers (Y = yes; N = no) their conspecific females have "decided."

burdened the female sex with a disproportionate responsibility for offspring care). Or, equivalently, anisogamy might be seen as the ultimate liberator of males with regard to many aspects of reproduction. However, with freedom of responsibility also comes a loss of control (and vice versa—with responsibility comes power). Almost by definition, male gametes normally are far more abundant than conspecific ova, so the latter are a limiting factor in any fertilization process that necessitates a union of the two. Moreover, because females produce and thereby control that limiting resource, the law of supply and demand implies that the female sex typically retains much of the ultimate authority over reproduction. Just how firm this feminine grip can be is illustrated by several reproductive quandaries (fig. 1.9) that primarily the female gender must resolve in various phylogenetic lineages. Several of these evolutionary "decision points" are elaborated in the sections that follow.

Provision Ova with Substantive Yolk?

For marine invertebrates with external fertilization and mobile pelagic larvae, biologists recognize two distinct ontogenetic trajectories based on the amount of yolk initially present in the egg. Lecithotrophic larvae get their nutrition from large yolks that represent a considerable maternal investment, whereas planktotrophic larvae develop from eggs with few or no food reserves (meaning that the larvae must feed while adrift in the ocean). Suites of life-history features go with these distinct ontogenies. For example, due to their endogenous yolk supplies, lecithotrophic larvae usually develop faster and have a shorter pelagic duration than their planktotrophic counterparts. They also tend to have simpler morphologies because they do not require complex feeding apparatuses. These differences in turn have had evolutionary ramifications ranging from species' clutch sizes to geographic distributions to rates of gene flow among conspecific populations (all of which tend to be greater in planktotrophic species) and even to speciation rates (which tend to be higher in lecithotrophic clades). For example, clutch sizes are typically smaller in lecithotrophic species because dams put more of their finite resources into yolks, leaving themselves with less energy to make additional eggs.

The term *lecithotrophy*, or "yolk feeding," is sometimes also applied to vertebrate animals when a female retains fertilized eggs within her body but offers no further internal nourishment to her embryos. Lecithotrophy then is distinguished from *matrotrophy*, or "mother feeding," in which a dam provisions her embryos more directly (as in mammalian pregnancies). In lecithotrophic species, the mass of an offspring at parturition is considerably lower than that of an egg at fertilization due to the gradual depletion of yolk as it supports the metabolic costs of embryonic development. In matrotrophic species, by contrast, an offspring at birth may weigh considerably more than the fertilized egg, evidencing the fact that the mother provided nutrients that more than offset the costs of embryonic maintenance. A somewhat analogous distinction applies to certain plants. In angiosperms (flowering plants), the maternal parent packages each zygote with a triploid endosperm that supplies the embryo with food immediately after the seed (which housed a zygote and an endosperm) germinates. An additional "weighty" maternal investment by angiosperms occurs when the dam encases her dormant seeds in a mature ovarian structure known as a fruit, which protects the seed and aids in its dispersal. In gymnosperms, by contrast, the dam packages her zygotes in "naked seeds" (albeit perhaps adorned with fleshy structures to facilitate dispersal) that contain haploid gametophytic nutritional tissue and are usually protected by woody cones.

Accept Sperm for Internal Fertilization?

As makers of ova and the gatekeepers to reproduction, females and their unfertilized eggs are sure to be pursued by conspecific males and their gametes. Thus, another evolutionary choice that anisogamy has relegated mostly to females pertains to the site of fertilization (inside versus outside the female body). For lineages in which females retain ova internally and then allow male gametes to enter through an orifice (e.g., via copulation), internal fertilization can obviously pave evolutionary paths toward female pregnancy and brooding. Furthermore, when the gametes from two or more males gain nearly simultaneous access to a female's reproductive tract, what routinely ensues are spirited sperm competition and cryptic female choice. Analogous opportunities for gametic competition arise in flowering plants when pollen grains from two or more donors arrive on a fertile female flower. Each grain then germinates on the stigma and begins to grow a pollen tube through elongate feminine tissue (the style) toward the flower's ovule-housed ova. In these cellular track meets or "pollen competitions," the fastest pollen-tube racers often achieve victory (successful fertilization), but the genetic composition and reproductive parts of female flowers help to decide the victors as well.

Encase Embryo in Egg?

For a phylogenetic lineage that has already evolved internal fertilization, the next key evolutionary choice for females is whether to encase each zygote and embryo in a yolk-rich egg. A "yes" answer can set the evolutionary course toward oviparity or egg laying (chapter 3), with further refinements in that decision influencing whether the eggs eventually become hard and impermeable (as in birds and reptiles) or softer and more permeable (as in fish and amphibians). In turn, the nature of the egg casing influences myriad ecological and evolutionary trajectories, not least of which is the type of habitat that a species might occupy (typically aquatic in the case of soft eggs and terrestrial in the case of hard eggs). However, when females have "decided" during evolution to retain naked embryos rather than encase them in shells, entirely different evolutionary trajectories (toward viviparity and pregnancy) might be launched.

Give Birth to Live Young?

Viviparity, or "live birth," entails the kind of pregnancy with which humans are most familiar. However, when viewed from the comparative perspectives of either animal development or evolution, viviparity is not as distinct from oviparity as it might at first seem. Instead, both phenomena reside along a continuum involving the many different times and places (relative to a "pregnant" individual's body) that pertain to the hatching of fertilized eggs (chapters 3 and 4).

Entrust Offspring Care to Males?

Finally, another evolutionary choice that females in various taxa have decided differently involves whether to entrust males with the exclusive care of ova and/or offspring. Such transfer of reproductive responsibility (and control) between the sexes is uncommon in the biological world, suggesting that this evolutionary step has not been taken lightly. Furthermore, unlike many other reproductive decisions during evolution, the outcome in this case is not simply "woman's prerogative" because males can be expected to have had a say in any such shift of the reproductive burden. At the very least, we might expect males in "sex-role-reversed" lineages to demand assurance of genetic paternity for the offspring they tend. Indeed, in pipefishes and seahorses, each pregnant male can be certain that he sired his brood because he fertilized the ova inside his own brood pouch.

Nevertheless, in many other animal species (such as nest-tending fishes and birds), males sometimes invest heavily in offspring care even without guarantees of paternity. For this and other reasons, "external male pregnancy" presents some interesting evolutionary departures from both standard female pregnancy and internal male pregnancy (chapters 7 and 8).

Other Sexual Asymmetries Related to Biological Parenthood

As described earlier, pregnancy is just one of anisogamy's evolutionary consequences that differentially affects the two genders. In addition, pregnancy in turn generates several other fundamental sexual asymmetries related to parenthood, including the assurance of maternity and the assessment of paternity.

Assurance of Maternity

Because of ova retention, the pregnant sex (typically the female) normally has complete assurance that she is the biological mother of all offspring in her brood. By contrast, her mate has no comparable guarantee that he is the genetic sire of those progeny (some or all of whom might have been fathered by other males). As elaborated in chapters 7 and 8, this huge difference between the sexes in the assurance of genetic parenthood has profound implications for the selective pressures that influence the evolution of mating systems, reproductive behaviors, and modes of parental care in species that display the pregnancy phenomenon.

Assessment of Paternity

A second sexual asymmetry that follows from the first is more technical than biological. For logistical reasons, molecular parentage analyses are usually best suited for assessing rates of multiple paternity within the broods of female-pregnant species and multiple maternity within the broods of male-pregnant species (DeWoody et al. 2000; Neff 2001). In other words, for any pregnant species in which the brooding parent can be collected with its offspring, geneticists can deduce the number of mates and frequencies of multiple mating by the brooders much more readily than they can for members of the nonpregnant sex. This asymmetry reflects the fact that each brood of half-sib embryos is physically associated with its pregnant sire or pregnant dam. By contrast, documenting the frequency of multiple mating by members of the nonpregnant sex is much more difficult because each such individual might have parented other broods not included in the genetic assays. This consideration becomes important when using molecular markers to assess the potential impact of pregnancy on the evolution of animal mating systems, mating behaviors, and sexual selection (chapters 7 and 8).

Evolutionary Reconstructions

Scattered throughout this book are examples of researchers' attempts to reconstruct evolutionary sequences of events that underlie various pregnancy-related features and phenomena. Such reconstructions are part of the "comparative method" (Harvey and Pagel 1991), wherein alternative expressions of a trait are mapped across related species in efforts to uncover the chain of historical events that led to the modern-day distributions of those traits. This approach, known more specifically as "phylogenetic character mapping," or PCM (Avise 2006), is outlined in box 1.6.

BOX 1.6   Phylogenetic Character Mapping (PCM)

Phylogenetic character mapping entails deducing the evolutionary histories of phenotypic characters (such as different fertilization modes or alternative categories of pregnancy) in a particular taxonomic group. A popular approach is to use an independent molecular estimate of phylogeny as a historical backdrop to map or "reconstruct" the likely route(s) of evolutionary transformation among those phenotypic traits. The process involves four steps: (1) gather molecular data (typically DNA sequences) from homologous genes in living species; (2) apply suitable phylogenetic algorithms to estimate a molecular phylogeny for those species; (3) survey and characterize relevant species for the phenotypes of interest (such as internal versus external fertilization); and (4) use suitable phylogenetic methods to deduce and plot probable character states throughout the tree. From such PCM exercises, robust conclusions about the evolutionary histories of particular phenotypes can often be drawn (Garland et al. 2005).

The basic concept of PCM is outlined in the accompanying figure (from Avise 2006). Shown across the top are eight hypothetical species (A–H) that display one or the other of two different character states of a phenotype, such as viviparity (black squares) versus oviparity (white squares) as alternative reproductive tactics. Knowledge of the evolutionary relationships of these eight species (e.g., from DNA sequences) can be used to determine how these reproductive modes evolved. For example, if species A–H prove to be phylogenetically related, as shown in diagram I, then oviparity was probably the group's original ancestral condition, and viviparity is a shared-derived condition (i.e., a synapomorphy) for the ADE clade. However, if the species are allied as shown in diagram II, then viviparity was the likely ancestral condition, and oviparity is a shared-derived state for the clade BCFGH. Many other outcomes are possible. For example, if the true phylogeny for the eight species is as shown in diagram III, then oviparity was probably the ancestral state from which viviparity evolved independently on three separate occasions, and the conclusion thus would be that viviparity was polyphyletic.

(*continued*)

BOX 1.6   (*continued*)

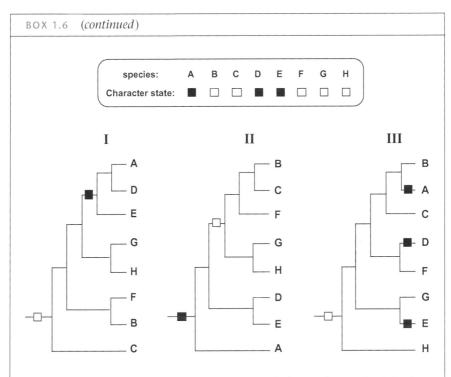

Phylogenetic character mapping is straightforward in principle, but complications and limitations often arise in practice. These raise problematic questions such as these: How accurate is the molecular tree? Were the phenotypes properly described? Were the phylogenetic reconstructions properly conducted? Thus, PCM-based conclusions (like all findings in science) should be viewed as informed but provisional.

SUMMARY

1. The reproductive life cycle in any sexual species is an endless succession of gamete production (gametogenesis), gamete deployment, gamete union (syngamy), and development (ontogeny) of the individual from zygote to adulthood. Because these four stages are sequential, where and when each takes place in a given species influences the trajectory of successive stages. In certain species, a particular form of pregnancy superimposes on this intergenerational reproductive process as an important component of ontogeny for both parent and child.

2. Gestational modes in various species can be matters of degree that defy attempts to define pregnancy by any single universal criterion. Among the many pregnancy-related biological features that collectively vary along continua are the following: the site of fertilization and early ontogeny (from intraparent to extraparent); the magnitude and nature of embryonic provisioning (from pure yolk feeding to extensive dam feeding); the category of parturition from the parent (from egg laying to live bearing); the magnitude of parental investment in progeny (from trivial to huge); and the precise site of parental care (from within-body to on-body to off-body). When defined broadly, pregnancy-like gestational phenomena in vertebrate and invertebrate animals offer a smorgasbord for comparative evolutionary analyses.

3. Typically associated with pregnancy are several landmark events, including gametogenesis, copulation, conception, implantation, and parturition, which not only help to distinguish alternative reproductive modes but also demarcate biological arenas within which different forms of selection take place. For example, the ontogenetic domain flanked by conception and birth, constitutes a temporal window within which viability-based natural selection is likely to materialize due to the intimate interactions between parent and fetus; alternatively, the arena flanked by copulation and syngamy offers fertile opportunities for intrafemale sperm competition and cryptic female choice. Evolutionary shifts in the timing of these and other ontogenetic events (heterochrony) can both shape and be shaped by natural selection and sexual selection.

4. Although mammals may embody the essence of pregnancy, many other animals display gestational modes that qualify as pregnancy-like by particular definitional standards. Even viviparous pregnancies have arisen in various nonmammalian taxa by several alternative and often surprising evolutionary routes, such as internal gestation following external fertilization in gastric-brooding frogs, and virgin births via parthenogenesis in several piscine and reptilian groups.

5. For better or for worse, gender-based asymmetries pervade reproductive operations in sexual species. Many of these sexual biases can ultimately be traced to an ancient evolutionary condition known as anisogamy: the strongly bimodal distribution of gamete size and mobility that continues to characterize essentially all sexual species. By definition, individuals who produce the relatively small and motile class of gametes are males, whereas those who produce the bulkier and less mobile class of gametes are females. Internal pregnancy itself, which typically burdens females far more than males, can be viewed both as a logical evolutionary outgrowth of anisogamy and as a profound amplifier of anisogamy's evolutionary effects. For example, preg-

nancy often makes the female sex even more of a limiting resource in reproduction and thereby amplifies the typical impact of natural selection and sexual selection on the two genders.

6. From an evolutionary perspective, anisogamy might be deemed the primal source of any female enslavement to procreation. On the other hand, anisogamy might equally be deemed to have been the ultimate liberator of males with regard to many aspects of reproduction. However, with freedom of responsibility also comes a loss of control (and vice versa: with female responsibility comes female power). During phylogeny, the female sex in effect has driven most of the key evolutionary decisions with regard to the general reproductive modes displayed by various taxa. Among the evolutionary "decision points" that have been basically relegated to the female sex are whether to provision the ova with yolk, accept sperm for internal fertilization, encase each embryo in an egg, give birth to live young, and entrust any postparturitional care of offspring to males.

7. Pregnancy also entails other types of sexual asymmetries. One biological example is the far greater assurance of genetic parenthood of a brood for females than for males, a fact that has many evolutionary ramifications. Another important consideration related to pregnancy is that geneticists can use molecular markers to deduce the incidence of multiple mating by members of the brooding sex much more readily than they can for members of the nonpregnant gender.

CHAPTER TWO

Vertebrate Live-Bearers: The Borne and the Born

This chapter introduces "live-bearing" vertebrates as ambassadors of pregnancy and as exemplars of diverse evolutionary topics that the pregnancy syndrome motivates. In other words, it describes vertebrate creatures in which progeny are borne internally by their mothers before being delivered as free-living beings to the outside world. This overview should prepare readers for later chapters that focus on other manifestations of pregnancy-like phenomena in both vertebrates and invertebrates.

Viviparity

Viviparity (from "vivi," meaning "alive," and "parity," meaning "borne") refers to the gestation and subsequent delivery of offspring from within the body of a biological parent. Viviparous species, including *Homo sapiens* and most but not all other mammals, offer quintessential examples of the pregnancy phenomenon. However, female mammals are not the only creatures that become pregnant. Except for birds (class Aves), all major vertebrate groups, including fishes, reptiles, and amphibians, contain at least some live-bearing species. Pregnancies in these taxa, like those in many viviparous invertebrates, otherwise vary in many features, such as the precise site of internal fertilization, the location and duration of embryonic incubation, the size of a brood, the number of genetically involved parents (as when a brood has multiple sires), the mode and magnitude of nutrient passage from adult to offspring, and even the gender of the gestating parent.

Viviparity in any animal species has some obvious evolutionary advantages, including the amelioration of direct predation on eggs and a buffering of embryos against environmental extremes (Packard et al. 1977; Bull and Shine 1979). However, viviparity also has some clear evolutionary downsides, including (a) a typical reduction in fecundity of the pregnant parent (compared to species without pregnancy); (b) survival costs to the adult gestator due to locomotory and other challenges (Miles et al. 2000; Ghalambor et al. 2004; Plaut 2002); (c) the fact that the death of a pregnant parent necessarily results in the loss of an entire brood; and (d) inevitable evolutionary conflicts of interest that arise between a gestating embryo and the parent that houses it (chapter 6). Thus, it is an oversimplification at best to view internal gestation as an evolutionary panacea for rearing offspring because many fitness costs and benefits must be considered (Trexler and DeAngelis 2003).

Evolutionary Origins

Among vertebrate animals (creatures with backbones), viviparity made its first evolutionary appearance in fishes, but the phenomenon has arisen time and again not only in a wide variety of distantly related piscine lineages (Gross and Sargent 1985; Mank et al. 2005) but also in many other vertebrate groups (Hogarth 1976; Blackburn 1992). Utilizing phylogenetic considerations, Blackburn (2005) tallied more than 140 independent evolutionary origins for viviparity among the backboned animals alone and concluded (p. 296) that "The evolution of viviparity may represent the most striking example of convergent evolution found among vertebrates."

Each convergent or parallel evolutionary transition from oviparity (egg laying) to viviparity typically necessitates profound shifts in ontogeny by both parents and offspring. For example, a common chain of events during this evolutionary transformation might entail the following series of biological innovations (Amoroso 1960; Wourms 1981): (a) alterations of both male and female reproductive tracts (as well as mating behaviors) in ways that promote internal as opposed to external fertilization; (b) a decrease in the proclivity of females to shed eggs (or perhaps an increase in their tendency to shed fewer eggs) (Rosen 1962); and (c) reductions in a reliance on the egg yolk for nourishing embryos and a concomitant elaboration of specialized structures (such as a placenta) and physiological mechanisms for a more direct transfer of nutrients from parent to fetus. Furthermore, all of these changes, especially the emergence of an intimate maternal-fetal relationship, pose selective challenges for biotic systems ranging from the morphological to the trophic and also for biological functions related to osmoregulation, excretion, respiration, circulation, endrocrinology, and immunology (Medawar 1953; Matthews 1955). In short, each evolutionary transition from oviparity to viviparity (or vice versa) can be a major biological undertaking.

Developmental Sequences

As characterized by Turner (1947) and Wourms (1981), ontogeny in viviparous fishes and many other vertebrates includes four signature events often relevant to pregnancy: ovulation (O), the release of a mature oocyte (or sometimes an embryo) from its follicle or pit in the ovary; fertilization (F) or conception, the genetic union of egg and sperm; hatching (H), the escape of an embryo from any egg membranes or shells that envelop it; and parturition (P), the final exit of an embryo from its pregnant parent's body. By convention, P is called birth or delivery in viviparous species, and it is termed *extrusion* or *oviposition* in oviparous (egg-laying) species. Across species, the temporal sequence of these four events includes both fixed and variable elements. For example, hatching cannot precede fertilization, and parturition cannot precede ovulation. Otherwise, however, various evolutionary shuffles have occurred in the orders of these four ontogenetic mileposts, and, indeed, many

FIGURE 2.1   The tiger rockfish, *Sebastes nigrocinctus*, representing a taxonomic family (Scorpionidae) that includes ovoviviparous species (chapter 3) in which each pregnant female retains egg-encased embryos until close to the time of both hatching and parturition.

of the shifts help to define different reproductive modalities. For example, because viviparity normally entails internal fertilization, a common sequence of events in live-bearing fish is O-F-H-P, whereas the normal order in an egg-laying species with external fertilization is O-P-F-H. However, some temporal inversions in ontogeny are subtler or even counterintuitive. For example, fertilization in viviparous bony fishes typically occurs in ovarian follicles prior to ovulation, such that the overall sequence of events is F-O-H-P. Apart from evolutionary changes in the orders of these ontogenetic events, there are also several instances in which distinctions between the events themselves can become blurred. For example, hatching and parturition essentially coincide in some shark species in which each embryo retains its egg capsule throughout the gestational phase of its development within its mother. In addition, in live-bearing rockfishes of the genus *Sebastes* (fig. 2.1), embryos typically hatch as relatively undeveloped larvae, but whether they do so just before or just after parturition is not always clear from the evidence available for species in the wild.

Trophic Relationships Between Mother and Fetus

Another important consideration for all forms of pregnancy involves the mechanism(s) by which embryos receive nourishment from their gestating parent. The delivery routes and the magnitudes of nutrient transfer vary greatly among taxa. Most mammals and many viviparous fishes have a special device known as a *placenta*, which serves as a maternal-fetal conduit that delivers parental resources to embryos and removes fetal wastes. Placentation is an extreme and direct form of *matrotrophy*, or "mother feeding" of embryos. However, embryos in some other live-bearing fishes secure their nourishment by other means, such as *lecithotrophy* (exclusive reliance on egg yolk), or perhaps expressions of matrotrophy that entail limited or indirect nutritional support from the dam. For example, depending on the species, matrotrophy sometimes involves a placental analogue (Blackburn 1999b) or "pseudoplacenta" (Turner 1940b), or it may entail the direct absorption of maternal secretions or other assets from the mother's uterine or ovarian environment. Although lecithotrophy versus matrotrophy is a standard distinction in the reproductive literature, the line between these two trophic modalities often blurs, such as when embryonic gestation in a live-bearing species begins with yolk feeding but later proceeds to maternal feeding after the progeny have hatched internally (chapter 3). The line can also blur in a structural sense, such as in some fish species in which derivatives of the yolk sac participate in the direct absorption or transfer of nutrients from the mother's reproductive tract (Turner 1947).

Factoid: Did you know? Placenta-like connections between a pregnant parent and its progeny are common in cartilaginous fishes (Chondrichthyes) but, with a few notable exceptions, are relatively rare in most groups of bony fishes (Osteichthyes) (Blumer 1979; Wourms 1981).

With respect to the magnitude of maternal provisioning, different live-bearing vertebrates collectively show a full gradation from little or no supplementation beyond the yolk per se to extensive investments that enable the dramatic growth of embryos in utero. A standard way that scientists gauge the magnitude of maternal support during a pregnancy is to monitor weight changes or developmental progress in offspring between conception and parturition. In matrotrophic species, embryos typically gain much weight depending on the exact amount of maternal investment. In lecithotrophic species, by contrast, a substantial net loss of total organic weight (up to 50%) typically occurs during gestation as embryos gradually use up their finite yolk supplies.

The Cast of Viviparous Vertebrates

Although placental mammals may represent the epitome of pregnancy, they are hardly alone in exhibiting intraparent gestation of their young. To introduce the taxonomic scope of viviparous pregnancy and some of the many species associated with it, the following sections present the vertebrate animals in which a pregnant female gives birth to live offspring following a substantial period of embryonic gestation within the dam. Most of the viviparous vertebrates described here are obligate live-bearers. In addition, a few fish species that normally display external fertilization and oviparity show occasional or "facultative viviparity" when a female is sometimes fertilized internally and then retains at least some eggs until the embryos hatch within her body. Indeed, internal fertilization and facultative viviparity are two of the initial transitional stages that Rosen (1962) proposed as evolutionary antecedents to constitutive viviparity.

Bony Fishes

Among the approximately 18,000 extant species of bony fishes (superclass Osteichthyes), only slightly more than 500 (ca. 3%) display female pregnancy (Wourms 1981). These live-bearing species are distributed throughout more than 120 osteichthyan genera, representing at least 14 taxonomic families (table 2.1). (One additional family [Syngnathidae; Syngnathiformes] also has viviparous species, but the gestating gender in that special case is the male [chapter 7].) Clearly, female-pregnant viviparity in bony fishes is polyphyletic, having evolved independently on more than a dozen separate occasions in different osteichthyan lineages (Gross and Sargent 1985; Blackburn 2005; Mank et al. 2005; Mank and Avise 2006a, 2006b). Some appreciation of the diversity (and polyphyletic origins) of viviparity in extant bony fishes can be gained by inspecting figure 2.2, which shows how modes of parental care and fertilization relevant to live-bearing are taxonomically arranged across a phylogeny for these animals.

Evolutionary pathways to female pregnancy in bony fishes have also been deduced from these exercises in "phylogenetic character mapping," or PCM (Mank et al. 2005). As summarized in figure 2.3, the standard evolutionary

TABLE 2.1   Taxonomic distribution of live-bearing in extant fishes

Shown are 14 taxonomic families of bony fishes (class Osteichthyes) plus 11 orders containing more than 40 taxonomic families of cartilaginous fishes (class Chondrichthyes), which all include at least some live-bearing, female-pregnant species. For the many primary references that underlie these reports, see tables 1 and 2 in Blackburn 2005.

Osteichthyes	*Chondrichthyes*[1]
1. Latimeridae (coelacanths)	1. Hexanchiformes (2 families)
2. Zoarcidae (eelpouts)	2. Squaliformes (6 families)
3. Parabrotulidae (false brotulas)	3. Pristiophoriformes (1 family)
4. Bythitidae (brotulas)	4. Orectolobiformes (7 families)
5. Aphyonidae (deep-sea livebearers)	5. Lamniformes (7 families)
6. Hemirhamphidae (half-beaks)	6. Carcharhiniformes (8 families)
7. Goodeidae (splitfins)	7. Squatiniformes (1 family)
8. Anablepidae (four eyes)	8. Echinorhiniformes (1 family)
9. Poeciliidae (aquarium live-bearers)	9. Pristiformes (1 family)
10. Scorpaenidae (scorpionfish and rockfish)	10. Torpediniformes (2 families)[2]
11. Comephoridae (oilfish)	11. Myliobatiformes (10 families)[2]
12. Embiotocidae (surfperches)	
13. Clinidae (blennies)	
14. Labrisomidae (weed blennies)	

[1]Taxonomy from Nelson 2006.
[2]All members of this order (the rays) show aplacental viviparity (as do many sharks).

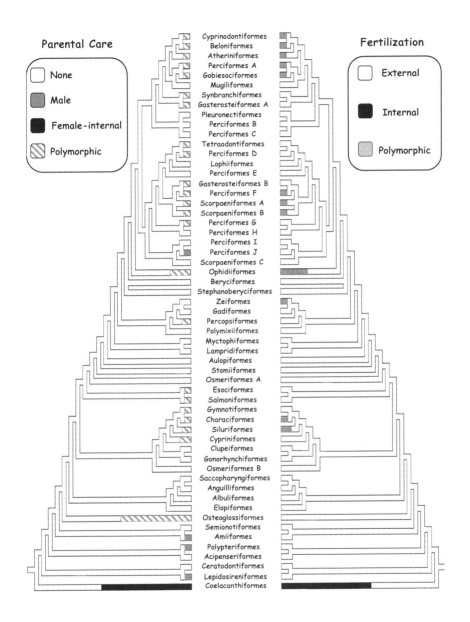

FIGURE 2.2 Ordinal-level phylogeny of Actinopterygii (a major subset of bony fishes) showing the phylogenetic distributions of alternative modes of fertilization and parental care (after Mank et al. 2005).

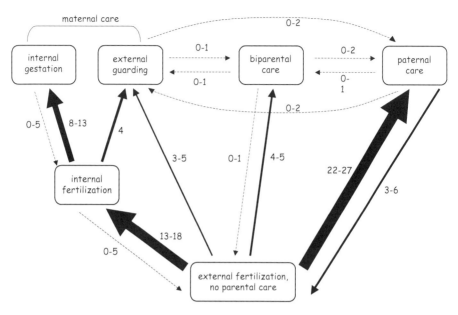

FIGURE 2.3   Evolutionary pathways to parental care in bony fishes as deduced from PCM analyses (after Mank et al. 2005). Arrow sizes reflect the relative numbers of transitions (numerals show minimum and maximum estimates) along each pathway. The left part of this figure shows the usual route to female pregnancy. The other forms of parental care in fish are discussed in chapter 7.

route proved to be as follows: external fertilization → internal fertilization → internal gestation (with few if any evolutionary transitions occurring in the opposite directions along this two-step pathway).

The following interesting details describe the better-known or most note-worthy families of extant osteichthyan fishes that include at least some species in which females become pregnant.

1. *Latimeridae* (*coelacanths*). As fossil evidence indicates, these prehistoric-looking fish, which were up to 6 feet long and had leglike fins, arose about 350 million years ago and were thought to have gone extinct about 65 million years ago, coincident with the asteroid impact that precipitated the demise of the dinosaurs. Then, in 1938, to everyone's amazement, a living coelacanth (*Latimeria chalumnae*) was trawled from deep waters off the eastern coast of South Africa. This specimen (fig. 2.4) and dozens more subsequently found in the Indian Ocean revealed many details about coelacanth biology that fossils could not readily provide. For example, researchers learned that female coel-

acanths are live-bearers that retain the developing young in their oviduct before delivering live-born young after a gestation that lasts somewhat more than a year (Smith et al. 1975). In the oviduct, the embryos seem not to be intimately connected to the dam but rather nestle unattached in cramped compartments, where they absorb oviductal secretions as food, perhaps supplemented by occasional oophagy (box 2.1).

BOX 2.1   Oophagy, Adelphophagy, and Histrotrophe

Apart from the more standard forms of lecithotrophy and matrotrophy, embryos in viviparous species also sometimes feed themselves by eating their siblings in utero. The term *oophagy* is the practice of embryos ingesting their parent's eggs during gestation. Many sharks produce "trophic eggs" apparently for this purpose. *Adelphophagy* (or embryonic cannibalism) is functionally similar but applies when gestating embryos eat one another rather than merely ingesting their mother's oocytes. Either or both of these phenomena are documented in various sharks (Compagno 1977; Stribling et al. 1980; Sims 2009) and certain other viviparous fish species (Wourms 1981; Veith 1980) and particular invertebrate animals (chapter 4) that either brood their offspring internally or encase them in egg capsules (Kamel et al. 2010). Adelphophagy is a form of cannibalism sensu stricto (Elgar and Crispi 1992) and can also be considered intradam siblicide or an extreme expression of sibling rivalry.

From an evolutionary perspective, oophagy and adelphophagy are not quite as macabre as they might at first appear. Oophagy can be interpreted as merely a modest exaggeration of the standard tendency of embryos in many viviparous fish and other species to ingest material known as *histrotrophe*: cellular debris and the decomposed products of moribund eggs and embryos that are the routine by-products of pregnancy in many live-bearing species, especially those with large broods. Oophagy often makes good sense for both the parent and its offspring because both parties are likely to profit in terms of personal genetic fitness when these extra resources in the mother's reproductive tract are put to good use rather than squandered. In turn, adelphophagy can be interpreted as a modest extension of oophagy to include the ingestion of live embryos. It, too, makes good genetic sense in some circumstances, such as when a pregnancy includes more offspring than could possibly survive to term, given the limited intrauterine space and other finite resources available inside a gestating parent.

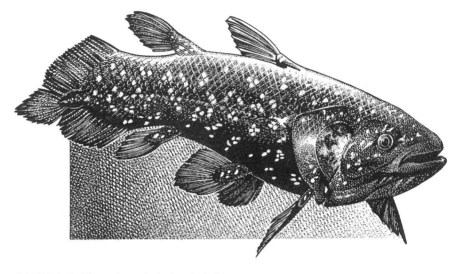

FIGURE 2.4   The coelacanth, *Latimeria chalumnae*.

Factoid: Did you know? Young coelacanths are called "pups," and there are typically 5–25 pups per litter. Some coelacanths may live to be more than 100 years old.

---

2. *Aphyonidae* (*deep-sea fish in the order Ophidiiformes*). These eel-like fishes (fig. 2.5) occur circumtropically in the bathypelagic realm, typically living at depths of 2,000–6,000 meters. Perhaps the most noteworthy features of their viviparous lifestyle relate to the fact that each male bundles his sperm into small sacs (*spermatophores*), which he delivers to a mate for prolonged yet viable storage inside her reproductive tract. This means that a male can mate with an immature female long before she has produced mature ova, yet his sperm may nevertheless fertilize eggs that she will produce after reaching sexual maturity much later in life. For these species, this special suite of reproductive behaviors (male sperm packing and female sperm storage) may be of crucial adaptive benefit to both sexes because conspecific mates might be hard to find in an environment characterized by perpetual darkness and perhaps low population densities.

3. *Goodeidae* (*splitfins*). In these small freshwater species (fig. 2.6) from Mexico and the southwestern United States, egg-encased embryos hatch

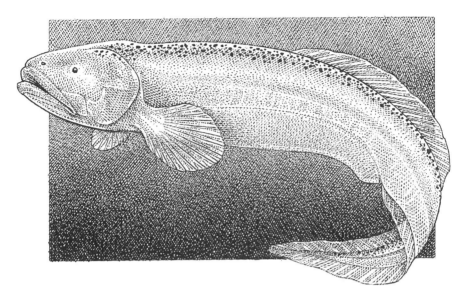

FIGURE 2.5   *Sciadonus cryptophthalmus*, a viviparous deep-sea fish in the family Aphyonidae.

inside their mother's body and then soon develop special placenta-like structures known as *trophotaeniae* (Turner 1937; Wourms and Lombardi 1979), which in this case are evolutionarily modified derivatives of the digestive system. The primary role of these trophotaeniae, which can be either rosette shaped or ribbonlike, appears to be to soak up nutritional resources from the ovarian environment and deliver them to the embryos housed therein until parturition. Furthermore, with respect to precise modes of ovarian gestation, there appears to have been an evolutionary trend in goodeid species toward increased specialization in embryonic feeding: from strict yolk feeding (lecithotrophy) on large-yolked eggs in some of the primitive goodeid lineages to extensive maternal feeding (matrotrophy) via elaborate trophotaeniae in many of the more advanced taxa (Turner 1933, 1940a; Lombardi and Wourms 1979).

Trophotaeniae also occur in several other live-bearing fish families, including some members of the Parabrotulidae, Zoarcidae, Clinidae, and Embiotocidae. After considering the structural diversity of piscine analogues of mammalian placentas, Wourms and Lombardi (1979) proposed that various groups of viviparous fishes convergently evolved these embryo-feeding adaptations by the following sequence of events: (a) origin of the trophotaeniae from a tubular embryonic

FIGURE 2.6   The redtail splitfin, *Xenotoca eiseni*, a live-bearing goodeid species with placenta-like structures.

gut; (b) precocious enlargement of the midgut or hindgut; (c) hypertrophy of the intestinal villi (fingerlike projections of the gut epithelium); (d) exteriorization of the gut lining and its villi; and (e) a further elaboration of trophotaenial rosettes and ribbons. The authors speculated that these evolutionary transitions took place via temporal shifts in the regulation of genes that govern the differentiation and ontogeny of intestinal cells.

4. *Poeciliidae* (*live-bearers*). This large Neotropical family includes mosquito fish (fig. 2.7) and many popular aquarium fishes (Jacobs 1971) such as guppies, mollies, platyfish, and swordtails. Although some of these viviparous species (such as the Amazon molly, pictured earlier in fig. 1.7) consist solely of females who reproduce by parthenogenesis or related clonal mechanisms (Avise 2008), most poeciliids are fully sexual. Nearly all of the more than 200 species in this family are viviparous, typically with gestational periods of about 1 month per brood (Breder and Rosen 1966; Meffe and Snelson 1989). Sperm storage by females is quite common in this group, as is the associated phenomenon of superfetation (box 2.2).

BOX 2.2   Superfetation

Superfetation, or *superembryonation*, is the simultaneous occurrence of two or more broods of embryos of different ontogenetic stages during a pregnancy. This condition is common in poeciliid fishes (Scrimshaw 1944a; Thibault and Schultz 1978), among whom 6 or more such cohorts may cohabit a female's reproductive tract (Scrimshaw 1944b). Superfetation occurs at least sporadically in several other viviparous fish assemblages as well (Turner 1947; Reznick et al. 2007), such as Zenarchopteridae (halfbeaks) and Clinidae (blennies), for whom as many as 12 broods per pregnancy have been reported (Veith 1979).

Superfetation normally requires either (a) sperm storage by females, in which case all cohorts carried by a pregnant female might have the same father, but different batches of eggs have been fertilized on various occasions and thus have matured at dissimilar times or (b) multiple mating by the dam, in which case some cohorts of offspring might have different sires. These requirements are not mutually exclusive, and females in many species of superfetatious poeciliids have both long-term sperm storage (lasting several months) (Vallowe 1953) and proclivities for mating with multiple males (Zane et al. 1999). For such taxa, genetic paternity analyses in conjunction with assessments of the ontogenetic stages of embryos can reveal how many sires contributed to each superfetatious pregnancy.

Superfetation probably arises in evolution when the periods required for oocyte maturation and/or embryonic development in a species shorten, thereby in effect permitting successive broods to be squeezed into the temporal window of a pregnancy (Turner 1947). In these situations new groups of oocytes can mature and be fertilized before other broods at more advanced stages of development have completed their gestation and undergo parturition. Such temporal adjustments in ontogeny are likely to be promoted by natural selection when they increase the reproductive outputs of pregnant dams and thereby, in effect, also happen to motivate transitions from *semelparity* (the occurrence of a single brood during an individual's lifetime) to *iteroparity* (repeated reproductive cycles within a lifetime) (Thibault and Schultz 1978). Superfetation seems to be best developed in species with refined mechanisms for nutrient transfer between a mother and her broods (Pollux et al. 2009), probably because matrotrophy improves the effectiveness of superfetation (Reznick and Miles 1989; Wourms 1981).

Male poeciliids have a modified anal fin known as a *gonopodium*, which serves as an intromittent organ (analogous to a penis) that each male wields with dexterity to inseminate females with special "sperm balls" called *spermozeugmata*. Each spherical spermozeugmatum is a collection of radially oriented sperm cells with their heads located peripherally and tails coiled into the interior. Brood size in a typical poeciliid pregnancy is 20–30 offspring.

As was also true for goodeid species, the maternal-fetal association in poeciliids ranges from strict lecithotrophy to refined matrotrophy (Reznick et al. 2002). Embryos in lecithotrophic species typically show a weight loss during gestation as they gradually use up

FIGURE 2.7   Mosquito fish (*Gambusia affinis*), a member of the live-bearing family Poeciliidae. Mosquito fish get their name from the fact that these animals eat aquatic mosquito larvae. Note that male mosquito fish (the two smaller specimens in this drawing) have a gonopodium (an anal fin with elongated rays), which acts as an intromittent organ during mating events with the larger females.

their finite yolks. Among the matrotrophic poeciliids, embryos in some species maintain a more or less constant gestational weight (indicating that they receive at least modest nutritional support from the mother), whereas in other species they show dramatic weight gains of perhaps 50-fold or more by the time of birth (thus indicating more extensive maternal support during gestation). These two categories of mother-fed species are called *unspecialized* and *specialized matrotrophs*, respectively. Perhaps not surprisingly, the magnitude of nutritional support that a poeciliid embryo receives from its dam is related to the nature of the physical connection between mother and fetus during each pregnancy. Some poeciliid species show little or no development of a placenta, whereas others show highly developed placenta-like structures that enable extensive maternal-fetal interchange (Panhuis et al. 2011), and still others show intermediate degrees of placentation that have helped researchers decipher the evolutionary transitions from aplacental to placental conditions (Pollux et al. 2009).

To explore the evolutionary history of maternal provisioning and placental development in poeciliid lineages, Reznick et al. (2002) used a molecular phylogeny as a backdrop. From this exercise in phylogenetic character mapping, the authors deduced that placental structures evolved on at least three separate occasions within the genus *Poeciliopsis* alone, plus several times elsewhere in the family (fig. 2.8). Thus, placenta-like structures and the associated phenomenon of matrotrophic feeding clearly have arisen by convergent evolution on multiple occasions in these fishes, a conclusion also reached by Pires et al. (2010). Taxa with well-developed placentas are often deeply embedded in clades that otherwise consist of species that mostly lack placentas and show little or no direct maternal provisioning of embryos.

5. *Embiotocidae (surfperches)*. More than 20 of these small live-bearing fish species (see fig. 2.9) inhabit mostly the shallow marine waters of the northern Pacific basin. Perhaps the most interesting feature of their advanced viviparity is that the reproductive life cycles of the two sexes are about 6 months out of phase (Wiebe 1968). In males, spermatogenesis begins in early spring and eventuates in mature sperm by June or July, at which time copulations take place following elaborate courtship displays. During a mating event, each male uses a system of appendages and tubular structures in his much-modified anal fin to transfer spermozeugmatic packets, each containing about 600 spermatozoa, to a female's reproductive tract. Afterward, his mate stores these gametes in special pockets of her ovarian epithelium. Oocyte maturation, followed by fertilization within the

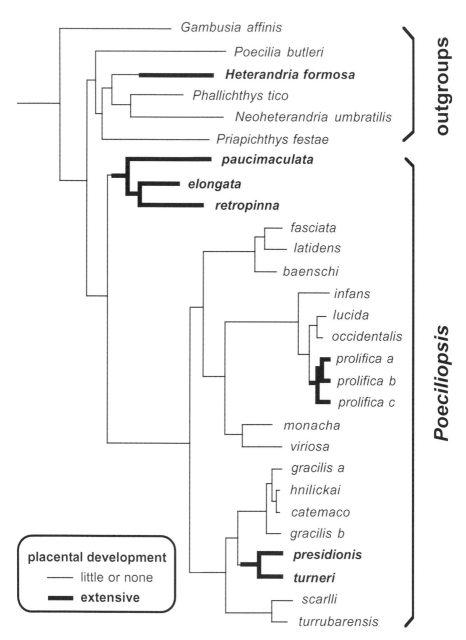

FIGURE 2.8 Molecular phylogeny for poeciliid taxa, demonstrating independent evolutionary origins of placentas in several of these piscine lineages (from Avise 2006, following Reznick et al. 2002).

FIGURE 2.9   The shiner perch, *Cymatogaster aggregata*, a live-bearing species representing the family Embiotocidae.

Factoid: Did you know? In another embiotocid species (*Micrometrus minimus*), newborn males are sexually active and inseminate newborn females, who then must grow up to become pregnant (Schultz 1993).

ovarian follicle, then takes place from October to December. Later, the nearly yolk-free eggs are released into the ovarian cavity, where the embryos gestate for about 6 months (Wiebe 1968), relying on nutrients supplied by copious secretions from their mother's ovigerous folds (Wourms 1981). Finally, the progeny are born in an almost adult condition (Baltz 1984).

Multiple mating and long-term sperm storage by embiotocid females undoubtedly contribute to the genetically documented fact that broods gestated by a pregnant female often include embryos from multiple sires (Reisser et al. 2009; Liu and Avise 2011).

6. *Bythitidae* (*viviparous brotulas*). These shallow-water inhabitants of tropical and subtropical reefs illustrate the wide range of brood sizes (i.e., fecundities) that sometimes characterize even closely related viviparous fishes. For example, whereas only about a dozen embryos

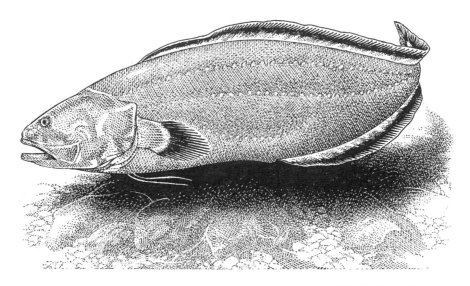

FIGURE 2.10   *Grammonus longhursti*, a live-bearing brotula fish (family Bythitidae) with extremely large broods.

comprise a pregnancy in members of the bathytid genera *Lucifuga* and *Ogilbia*, several hundred embryos are borne by pregnant dams in *Dinamatichthys iluocoeteoides*; moreover, in *Grammonus* (*Oligopus*) *longhursti* (fig. 2.10), a pregnant mother gives birth to as many as 15,000 progeny. Presumably, species that fall closer to the oviparous end of the ovoviviparity spectrum (see chapter 3) can have higher fecundities than strictly viviparous species, all else being equal, because offspring from "oviparous pregnancies" tend to be born at less advanced stages of development.

Cartilaginous Fishes

Unlike the bony fishes, among whom viviparity is relatively uncommon and taxonomically localized, about 60% of 970 extant cartilaginous species (mostly sharks and rays) give birth to live young. These female-pregnant live-bearers reside in approximately 130 genera in the superclass Chondrichthyes and in nearly 50 condrichthyan families (examples in table 2.1), representing nearly a dozen taxonomic orders (Wourms 1977, 1981). Typically, females produce small numbers of heavily yolked eggs enclosed in distinctive egg cases (fig. 2.11) that hatch either internally (in viviparous species) or externally (in oviparous species).

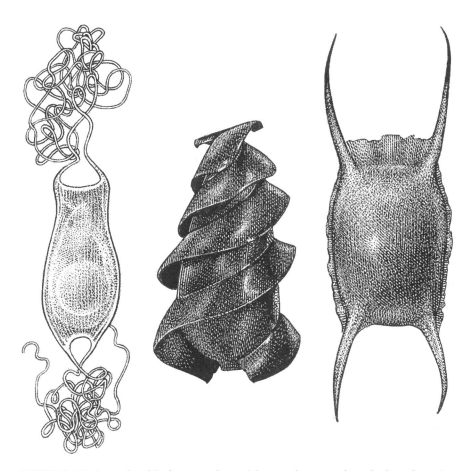

FIGURE 2.11   Examples of the beauty and varied designs of egg cases from sharks and rays. In some cartilaginous fish species, embryos hatch from such egg cases inside the mother and are born live, but in other species the embryos hatch after the egg cases have been deposited into the environment. From left to right: cat shark (*Apristurus* sp.), horn shark (*Heterodontus francisci*), and undulate ray (*Raja undulata*).

Placentation seems to be confined to the ground sharks (Carcharhini-formes), where it probably evolved just once or twice (Dulvy and Reynolds 1997). Thus, overall, sharks display a wide array of reproductive modes, including oviparity, aplacental viviparity, and placental viviparity (Byrne and Avise 2012).

internal fertilization, and pregnancy had already arisen in some vertebrate lineages a long time ago.

Amphibians

Among the world's approximately 6,000 extant species of frogs and toads (order Anura), plus 600 species of salamanders and newts (Urodela), only a small handful (less than 1%) can be considered strictly viviparous in the sense of giving birth to live offspring that were conceived and then "oviductally gestated" within their mothers (Taylor and Guttman 1977; Greven 1998; Wells 2007). Indeed, only one extant species of salamander (fig. 2.15) and just a few frog and toad species exhibit this refined gestational mode (Wake 1993). All of these viviparous species have internal fertilization, but the mechanics differ (Kühnel et al. 2010): in the salamander, a female becomes pregnant after she picks up a spermatophore from the substrate, whereas an apposition of male and female cloacas (mating) precedes pregnancy in most viviparous frogs and toads.

FIGURE 2.15   The alpine salamander, *Salamandra atra* (Salamandridae) of the Central and Eastern Alps is the only salamander species known to be an obligate live-bearer; a pregnancy may last 2–5 years.

Given the rarity of viviparity in most amphibians, it comes as quite a surprise that a majority of the approximately 190 extant species in a third amphibian order, Gymnophiona (the "caecilians"), routinely give birth to live young (Wake 1977a, 1977b, 1993). In these animals, which superficially look like worms (fig. 2.16), internal fertilization is achieved when a male everts the rear portion of his cloaca to form a phallodeum that he insets into the vent of a female. The embryos gestate internally (nourished by yolk and/or by a mother's oviductal secretions but fed also by a pseudoplacenta in a few species). By the time of birth, the young typically have metamorphosed. Both the size and the number of ova tend to be reduced in the viviparous (as compared to oviparous) caecilians, as is also generally true for the other live-bearing amphibians. Based on a molecular phylogenetic analysis, viviparity apparently evolved independently at least four times in the Gymnophiona (Gower et al. 2008).

FIGURE 2.16   A viviparous caecilian, *Gegeneophis seshachari*, the only known live-bearing amphibian from Asia.

FIGURE 2.18   An example of a live-bearing marine reptile, the elegant sea snake, *Hydrophis elegans*.

FIGURE 2.19   The Madagascar chameleon, *Chamaeleo oustaleti*. Although this and most other chameleons on the island of Madagascar are oviparous, a few South African members of this reptilian clade (such as the dwarf chameleon, *Bradypodion pumium*) have evolved viviparity.

Mammals

With just a few exceptions (the egg-laying monotremes described in chapter 3), all extant mammals are viviparous. Typically, these modern live-bearing species are categorized as members of either of two related "sister taxa"—the placentals (infraclass Eutheria; see fig. 2.20) and the marsupials (infraclass Metatheria). However, this taxonomic distinction can be somewhat misleading because all living Metatheria and Eutheria have a placenta that intimately connects mother and child during each intrauterine (i.e., intrawomb) pregnancy. Both groups also suckle their postpartum young. Another source of potential confusion is that although nearly all extant mammals are viviparous, many extinct mammalian species were not, and, indeed, "Most mammalian lineages in the history of the world were probably egg-layers" (Tudge 2000, 437), from which live-bearing might well have evolved more than once independently.

If both marsupials and "placental" mammals utilize a placenta, what, then (if anything), reproductively distinguishes the eutherians and the metatherians? Actually, it is mostly just a matter of differing emphases. In comparison

FIGURE 2.20   A representative menagerie of viviparous Eutherian mammals.

FIGURE 2.21   The western gray kangaroo, *Macropus fuliginosus* (Macropodidae), and musky rat kangaroo, *Hypsiprymnodon moschatus* (Hypsiprymnodontidae), two representative metatherian mammals or marsupials.

to the pregnancies in most marsupials, eutherian pregnancies tend to be longer and more refined such that many (but certainly not all) eutherian species deliver their young in quite an advanced state of development. Consider, for example, a foal, which is able to feed from the mare and frolic soon after parturition.

Marsupials (fig. 2.21), by contrast, normally give birth to less mature babies, even in species whose pregnancies can be quite long due to embryonic diapause (chapter 6). Also, many marsupial mothers compensate to some extent for giving "premature" births by carrying and suckling their newborns in special belly pouches.

SUMMARY

1. Viviparity is the bringing forth of offspring that are first gestated internally by a parent. With few exceptions, the sex of the gestating parent is female.

Except for birds (class Aves), all major vertebrate groups contain at least some viviparous (live-bearing) species. Pregnancies in these taxa vary in many features, such as the precise site of internal fertilization, the location and duration of embryonic incubation, brood size, the number of parents genetically involved (as when a brood has multiple sires), and the mode and magnitude of nutrient passage from adult to offspring.

2. In backboned animals, viviparity made its first evolutionary appearance in fishes nearly 400 million years ago, but the condition is highly polyphyletic, having arisen on more than 100 separate occasions in vertebrate lineages alone. Each evolutionary transition from oviparity (egg laying) to viviparity typically entails morphological and behavioral shifts toward internal fertilization, plus suitable adjustments between mother and embryo in biological functions related to nutrition, excretion, respiration, circulation, and endocrinology. Nearly all mammals and many viviparous fishes and reptiles have either placentas or placental analogues that serve as maternal-fetal conduits that help to carry out many of these physiological functions.

3. Viviparity is rather uncommon (3% of extant species) in bony fish (Osteichthyes) but predominates (70% of extant species) in cartilaginous fishes (Chondrichthyes). This disparity probably relates to the fact that viviparity requires internal fertilization, which is ubiquitous in extant cartilaginous fishes, whereas the vast majority of bony fishes fertilize ova externally. The site of pregnancy also differs between most viviparous chondrichthyans and osteichthyans; it is uterine in all live-bearing cartilaginous fishes but uniquely intraovarian in most viviparous bony fishes. With regard to embryonic sustenance within the dam, extant live-bearing fishes collectively show a wide diversity of nutritional modes ranging from strict lecithotrophy (yolk feeding only) to placental matrotrophy (advanced feeding by the mother) to embryonic cannibalism.

4. In the class Amphibia, viviparity is extremely rare in the taxonomic orders Anura (frogs and toads) and Urodela (salamanders and newts). However, live-bearing predominates in a third amphibian order—Gymnophiona (caecilians)—where the phenomenon is polyphyletic and entails several different mechanisms by which embryos receive nourishment while gestating within their pregnant mother.

5. Although most species in the class Reptilia lay eggs, more than 50 different reptilian groups contain at least a few live-bearing species that give birth to live young, thus indicating that viviparity in reptiles is highly polyphyletic. Nevertheless, all of these viviparous taxa reside in various lineages in the order Squamata (snakes and lizards), where females in about 20% of extant species give birth to free-living offspring following pregnancies of varied durations and different modes of embryonic gestation within the dam.

6. Extant live-bearing mammals are traditionally classified as either marsupials or placentals. However, pregnancies in both of these groups entail the intrawomb gestation of an embryo that receives nutrition via a placenta while in utero and suckles milk from its mother after birth. Thus, although marsupial females normally give birth to babies that are less mature than their placental counterparts (and then may carry their offspring in an external pouch), the different expressions of viviparity in these two mammalian groups are mostly matters of degree. Furthermore, when extinct mammalian lineages are taken into consideration, perhaps a majority of mammalian species that have inhabited the planet might well have been egg layers rather than live-bearers.

he grouped species into several binary morphological categories such as winged versus wingless and two legged versus four legged, and he also noted a fundamental distinction between species that give birth and those that lay eggs. In modern parlance, the former reproductive mode is termed *viviparity* (chapter 2), which biologists continue to contrast with oviparity, or egg laying, a major topic of this chapter. However, before proceeding to the lineup of oviparous vertebrates, several questions about the nature of the lain egg must be addressed.

What Kinds of Eggs Are Shed?

Oviparous females lay eggs (by definition), but the nature of the shed egg has a profound impact on each species' reproductive lifestyle. One key distinction is between fertilized eggs (containing a zygote or perhaps a multicellular embryo) and unfertilized eggs (ova or oocytes). When a female lays fertilized eggs, this implies that she was fertilized internally, either following a mating event involving direct insemination by a male or perhaps after acquiring sperm by some other means such as spermatophore capture. Conversely, when a female merely lays oocytes, the usual implication is that she belongs to a species with external fertilization. Although we can say that an ova-laying female was gravid, whether she qualifies as being pregnant is a matter of definition.

Interestingly, one sexual fish species has abandoned copulation entirely but sheds fertilized eggs nevertheless. It is the mangrove killifish (fig. 3.1), a hermaphroditic species in which each dual-sex individual simultaneously produces eggs and sperm that normally unite within the fish's body to yield zygotes that are then laid externally. *Kryptolebias marmoratus* and some of its close relatives are the only vertebrate animals known to reproduce via self-fertilization (an extreme version of inbreeding). They and other hermaphroditic species (Avise 2011) also provide a rather technical exception to the standard rule that only pure females can produce ova.

Factoid: Did you know? The mangrove killifish is actually an "androdioecious" species; a few pure males also exist in populations otherwise composed solely of hermaphroditic individuals. Furthermore, these males sometimes mediate outcross events when their sperm encounter unfertilized ova that the hermaphrodites occasionally release.

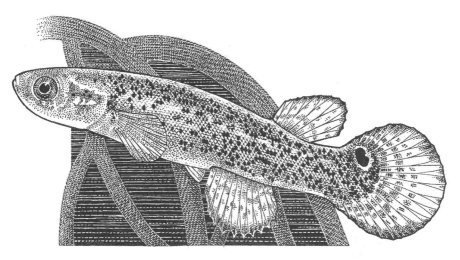

FIGURE 3.1   The mangrove killifish (*Kryptolebias marmoratus*), the world's only hermaphroditic vertebrate that regularly self-fertilizes.

What Is the State of the Offspring at Parturition?

A second key distinction relevant to the concept of "pregnancy" is whether the fertilized egg was hatched or unhatched at its time of shedding. In some species, each female carries fertilized eggs that hatch inside her body, whereas the hatching events in other species may be delayed until after parturition. Progeny are live-born in the former (viviparous) circumstance, whereas they are hatched externally in the latter (oviparous) situation, but the relative durations and degrees of embryonic development that take place within (rather than outside) the mother's body vary greatly across species. As described later, *ovoviviparous* taxa display varying elements of oviparity and viviparity in the sense that the intrafemale gestation of embryos occurs partly inside and partly outside an egg.

What Type of Casing Surrounds the Egg?

After an egg has been fertilized within a female, it seldom remains naked for long. Instead, in *amniotes* (all birds, reptiles, and mammals), each embryo soon becomes surrounded by several membranes: an *amnion*, which protects and suspends the embryo in a fluid-filled sac; a yolk sac, which stores nutrients; an *allantois*, which serves as a repository for some of the embryo's wastes; and a *chorion*, which collaborates with the allantois in gas exchange and embryonic respiration. Collectively, these membranes represent a major evolutionary advance over the reproductive apparatuses in nonamniotes, including all fishes and amphibians. Another ancient innovation during vertebrate evolution was the emergence of a *cleidoic*, or closed, egg, which, because of its waterproof casing or shell, is relatively resistant to desiccation and to some other kinds of environmental insults. The cleidoic eggs of reptiles and birds are very different from their gelatinous predecessors, which continue to confine reproduction in most fishes and amphibians to aqueous environments. Indeed, the initial evolution of the cleidoic egg, several hundred million years ago, played a key role in enabling vertebrates to colonize the terrestrial world. Interestingly, mammals still retain the four extraembryonic membranes, but most extant mammalian species have jettisoned the cleidoic egg in favor of maternal protection and internal gestation of their embryos via a placenta.

The eggshell secreted by the oviduct of any female vertebrate can serve several functions: protect the embryo from physical damage; permit the exchange of respiratory gases necessary for life; and perhaps provide an additional source of nutrition for the embryo. Sometimes these functions come into opposition, producing evolutionary trade-offs. For example, whereas a thick and highly calcified eggshell offers greater protection and may provide additional nutrients to a developing embryo, it also lowers porosity and generally reduces the efficiency of gaseous transfer when compared to an egg with a much thinner covering. Such functional trade-offs must be resolved somehow in any evolutionary transition between different modes of egg production and deployment (Stewart et al. 2010).

The Cast of Oviparous Vertebrates

All major vertebrate groups, including bony fishes, cartilaginous fishes, reptiles, amphibians, and mammals, contain at least some oviparous species, as do countless invertebrate taxa. However, oviparity is ubiquitous in only one vertebrate class (Aves), so it is with birds that we begin our survey of oviparous vertebrates.

Birds

In all 10,000 species of extant birds, each embryo is encased in a hard-shelled cleidoic egg that hatches only after exiting the mother's body. A bird's egg is an impressive contraption, typically tapered at one end so that it cannot easily roll out of a nest, large enough to provision an embryo, yet small enough to pass through a female's vaginal opening (cloaca), and protected by a hard casing that withstands the weight of incubating parents yet fragile enough to permit a hatchling to peck its way free. Furthermore, housed inside each cleidoic egg is everything necessary to support a developing embryo, including a nutrient-rich yolk, a layer of albumin, which supplies water and serves as a shock absorber, an allantoic sac, which helps the embryo respire and acts as a septic tank for waste products, and other specialized membranes that surround and separate all of the above. In short, the avian egg is a beautifully self-contained, out-of-body incubator.

The genesis of an egg begins deep within a hen (or, we might say, the genesis of a hen lies deep within an egg). Let us begin with the hen. In her upper reproductive tract is an ovary filled with oocytes at various stages of maturation. During the breeding season, as each ovum matures in its follicle, it swells and acquires a yolk. Then, during ovulation, the mature ova are released one at a time into the *infundibulum* (upper portion of the oviduct), where fertilization may take place if the hen has recently mated. Each fertilized egg (henceforth, an embryo as the zygote begins to divide) then begins its journey down the oviduct, first encountering the *magnum* region, where it settles for about an hour while albumin is incorporated. The embryo then moves to

FIGURE 3.2   Hen, chick, and cleidoic eggshell.

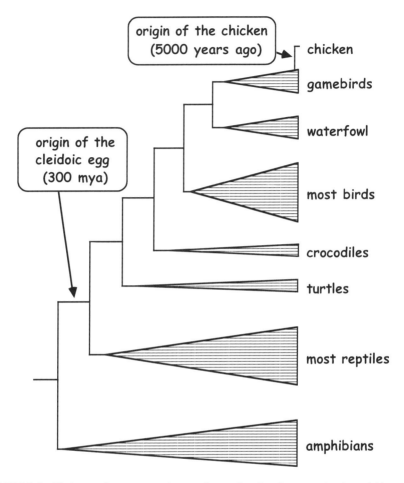

FIGURE 3.3   Phylogeny for representative vertebrates showing the approximate evolutionary ages of chickens vis-à-vis cleidoic eggs (from Avise 2006).

the *isthmus* section of the oviduct, where it resides for an hour as egg membranes are added. Next comes the *uterine* portion of the oviduct, where the embryo and its associated paraphernalia become encased in a carbonate shell, a process that takes several hours. Finally, the hen lays her egg, which later hatches an independent chick (fig. 3.2).

From an evolutionary perspective, eggs are a hen's way of making more chickens, and chickens are an egg's way of making more eggs. But which came first? With respect to an individual's development (ontogeny), this proverbial question has no answer because eggs and hens are just alternating phases of the avian life cycle. However, we can now declare (with tongue only partly in cheek) that eggs came long before chickens, and here is the phylogenetic reason. As summarized in figure 3.3, the domestic chicken (*Gallus domesticus*) is descended from red junglefowl (*Gallus gallus*), which people domesticated about 5,000 years ago, probably in India. Also producing cleidoic eggs are all other birds, plus the reptilian lineages from which ancestral birds separated several hundred million years ago. Phylogenetically speaking, birds are feathered reptiles, one subset of a far more ancient reptilian clade. Thus, the cleidoic egg arose approximately 300 million years ago, making it vastly older than the chicken. Clearly, the cleidoic egg came first!

Bony Fishes

In approximately 97% of all extant species of bony fishes (Osteichthyes), sexually mature individuals spawn oocytes (unfertilized eggs) that typically undergo syngamy in the external environment. As in other oviparous vertebrate and invertebrate species, spawning females are normally considered to be gravid but not necessarily pregnant. After the oocytes have been fertilized and have exited the mother's body, their fates vary tremendously, depending on the taxon. Broadcast spawners normally abandon their young, but in many other species the sire (or occasionally the dam) tends the embryos and the fry, typically in nests or other special brooding sites. As detailed in chapters 7 and 8, it can sometimes be useful to think of such extended care-giving species as displaying "external pregnancy."

Cartilaginous Fishes

Unlike in bony fishes, where oviparity is extremely common and taxonomically widespread, oviparity in the cartilaginous fishes (Chondrichthyes) is confined mostly to a few taxonomic families of sharks, skates, and chimaeras. An example of the latter is the monstrous Chimaera (fig. 3.4), a deepwater fish sometimes dredged from the North Sea. In this species, internal fertilization follows copulation, which is probably facilitated by a forehead clasper that each male uses to hold his mate. An impregnated female then houses the resulting embryos in half-foot-long egg capsules, which hatch after she has laid them outside her body.

Despite the relatively high frequency of viviparity in sharks and rays (chapter 2), oviparity is almost certainly the ancestral reproductive mode from which live-bearing has evolved independently several times in this clade of cartilaginous fishes (Dulvy and Reynolds 1997).

Amphibians

Except for the viviparous caecilians described in chapter 2, nearly all female amphibians lay their eggs externally (typically in water) before they are

FIGURE 3.4   *Chimaera monstrosa* (Chimaeriformes), an example of an oviparous cartilaginous fish.

fertilized. However, with respect to any subsequent parental care of the zygotes or young, what happens next in these oviparous animals varies greatly from species to species (Duellman and Trueb 1986; Wake 1993). Most parents just walk, hop, or swim away, but some show more attention to their offspring. For example, some frogs make foam nests, some actively transport fertilized eggs to nearby water, and some even deliver food to their developing progeny (Weygoldt 1980). In the so-called marsupial frogs (fig. 3.5), mothers are even more attentive: they carry their developing progeny in brood pouches (special skin folds) on their backs (Jones et al. 1973). Here is how this happens: A male catches a female's unfertilized eggs as she lays them; then he deposits

FIGURE 3.5   The Andean marsupial frog, *Gastrotheca riobambae*, a species in which mothers carry offspring in a pouched "backpack."

and fertilizes these ova in the dam's "backpack," or dorsal brood pouch, where she protects her babies until they hatch and drop into the water to begin independent lives.

Reptiles

As precursors to the birds (and mammals), reptiles were the first vertebrates to deploy cleidoic eggs and thereby free themselves from the need to reproduce in an aqueous environment. Thus, although many snake and lizard lineages include species that give birth to live young (chapter 2), oviparity was clearly the ancestral (and remains the most common) mode of reptilian reproduction. In squamate reptiles, the embryos of most oviparous species are lecithotrophic and thus receive their nourishment primarily from yolk (albeit supplemented with calcium from the eggshell). Viviparous squamates are mostly lecithotrophic, too, but they also show various degrees of matrotrophy (*placentotrophy*). From these and other observations, Stewart and Thompson (2000) argue that the pattern of embryonic nutrition in oviparous snakes and lizards is probably an *exaptation* (biological precondition) for the evolution of viviparity in this taxonomic assemblage.

Mammals

Except for five living species of *monotremes* (echidnas and platypuses), all extant mammals are viviparous. The semiaquatic duck-billed platypus (fig. 3.6) is a bizarre Australian creature with many odd features such as a rubbery ducklike bill, a flattened beaverlike tail, otterlike webbed feet, a venomous spur on the male's hind foot, and the distinctly unmammal-like habit of oviparity. During the breeding season, a female lays 1–3 small leathery eggs following a month of internal gestation, which precedes another 10 days of external incubation before her eggs finally hatch. Reproduction in Australia's short-beaked echidna (fig. 3.7) is generally similar, but a female typically lays only 1 egg per year.

Monotremes get their name from the Greek "monos" (one) and "trema" (hole), which refers to the single cloacal opening through which a female urinates, defecates, and lays her fertilized eggs. (Marsupials have a separate opening for the genital tract, and female placental mammals have three such openings—one each for their reproductive, excretory, and digestive systems.) Although monotremes appear to differ rather fundamentally from marsupial and placental mammals by virtue of being oviparous, the mother retains the fertilized eggs internally and supplies them with nutrients for some time, so in that respect each monotreme female exhibits several elements of standard mammalian pregnancies. She also lactates and thereby offers prolonged support to her hatchlings (known as "puggles"). Extant monotremes are thought to be the descendants of an early offshoot of the mammalian evolutionary tree.

Factoid: Did you know? When European naturalists first encountered the platypus in Australia in 1798 and sent dried skins back to England, British scientists at first thought it might be the hoax of a taxidermist who had sewn together body parts of unrelated animals such as a duck and a beaver.

Factoid: Did you know? Echidnas are also called spiny anteaters because their bodies are covered in spines and they eat ants.

C
C
f

FIGURE 3.6   The duck-billed platypus, *Ornithorhynchus anatinus*, an egg-laying monotreme mammal.

Factoid: Did you know? Although female platypuses and echidnas have mammary glands that secrete nutrient-rich milk, they lack teats. Instead, their milk is released through pores in the skin and then pools in abdominal grooves where it is lapped up by their offspring.

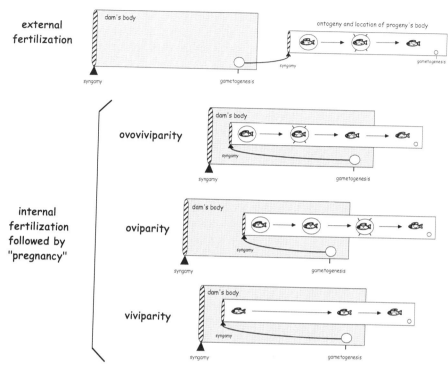

FIGURE 3.8   Diagrams of viviparity, oviparity, and ovoviviparity. Note that even simple spatiotemporal shifts in juvenile ontogeny (vis-à-vis the dam) might account for evolutionary interconversions among these three gestational modes. This can be seen from the diagram by imagining that the boxes representing generation 2 (the developmental timeline of progeny) shift either to the left or to the right with respect to generation 1 (the body of a mother) during the evolutionary process.

grow inside their mother until they are born, when they are about 50 cm long. The frilled shark appears to be aplacental (without a placenta), but the mechanism by which gestating embryos receive nutrients from their pregnant dam remains to be determined.

Factoid: Did you know? Some sharks and other live-bearing fishes produce what in effect are "trophic eggs," which are eaten by developing embryos within the mother prior to hatching (Heemstra and Greenwood 1992). Perhaps even more remarkably, females in one species of catfish reportedly extrude unfertilized eggs to help feed hungry fry in their nests (Helfman et al. 1997, 360).

FIGURE 3.9   The frilled shark, *Chlamydoselachus anguineus* (Chlamydoselachidae), an ovoviviparous deep-sea shark.

FIGURE 3.10   The Atlantic spiny dogfish, *Squalus acanthias* (Squalidae), an ovoviviparous shark that is mostly lecithotrophic (yolk-fed).

Factoid: Did you know? The spiny dogfish used to be one of the world's most abundant shark species, but overfishing for pet food, fertilizers, liver oil, and dissection specimens for biology classes have led to its recent listing as critically endangered in the northeastern Atlantic Ocean.

Another example of an ovoviviparous species is the Atlantic spiny dogfish, *Squalus acanthias* (fig. 3.10) (Squalidae). In this cartilaginous species, several embryos are housed in an elongated egg case that remains intact for half a year before hatching internally, after which the hatchlings gestate in their mother's uterus for another year or more, apparently without receiving much more nutritional support from the dam.

Evolutionary Conversions

In chapter 2, we noted that although viviparity is much less common than oviparity in bony fishes, live-bearing and its associated reproductive conditions (including internal fertilization and embryonic gestation within the dam) clearly are polyphyletic, having arisen on at least several independent occasions in these lineages (see figs. 2.2 and 2.3). To address such evolutionary transitions in greater depth, Mank et al. (2005) employed a "supertree" for ray-finned fishes as a phylogenetic backdrop against which to tally the deduced numbers of evolutionary switches among alternative reproductive modes and among various methods of parental care (which also include the external care of embryos by at least one parent in the many bony fishes with external fertilization).

The results of this exercise in phylogenetic character mapping (PCM) yielded two major conclusions: (a) internal fertilization arose from external fertilization on more than a dozen occasions in these fishes, whereas evolutionary switches in the reverse direction (internal → external fertilization) were rare at best, and (b) syngamy inside the female body (internal fertilization) was the usual stepping-stone on the evolutionary pathway toward internal gestation (conventional female pregnancy) in these piscine lineages.

Dulvy and Reynolds (1997) used similar PCM approaches to address evolutionary transitions between oviparity and viviparity in a group of cartilaginous fishes that includes sharks and rays. Figure 3.11 summarizes the authors' interpretations of their exercise in PCM, which indicated the following: (A) live-bearing evolved from egg laying on nearly a dozen occasions in these fishes, whereas changes in the opposite direction were rare, and (B) transitions among various types of maternal provisioning (along the scale from lecithotrophy to matrotrophy) were less consistent in magnitude and

(A)

(B)

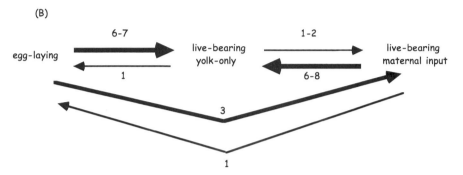

FIGURE 3.11   Phylogenetically deduced numbers of evolutionary transitions among various reproductive modes in the sharks and rays (after Dulvy and Reynolds 1997). (A) transitions between egg laying and live-bearing; (B) transitions among additional parental-care modalities. Arrow widths are proportional to the deduced numbers of evolutionary transitions along the pathways.

direction, often showing evolutionary reversals. Although some of these conclusions remain provisional (see Blackburn 2005), the findings again demonstrate the general evolutionary fluidity of interconversions between oviparity and viviparity in fishes.

Traditionally, biologists often asserted that viviparity in vertebrates can evolve rather readily from oviparity but that changes in the reverse direction have been rare to nonexistent (but see de Fraipont et al. [1996] for an opposing view and see Lynch and Wagner [2010] for a contrasting empirical example involving boa snakes). The basis for the conventional viewpoint seemed to be that producing an eggshell plus all of the other structures associated with oviparity presumably requires special metabolic pathways and organ systems that could not easily be regained once lost during evolution (box 3.1). However, the evolution of viviparity in any vertebrate lineage would also seem to involve the acquisition of numerous complex adaptations (Amoroso 1968; Schindler and Hamlett 1993), such as for fetal respiration and nutrition inside the female's body and the avoidance of maternal-fetal physiological incompatibilities. Thus, it is not altogether clear which is more complex: viviparity or oviparity. Furthermore, both egg laying and live-bearing have advantages and disadvantages that probably depend at least in part on ecological circumstances, such that neither of these reproductive strategies is likely to be universally superior to the other in all taxa or in all environmental settings. Thus, natural selection might be expected to push different species in different evolutionary directions along the oviparity-to-viviparity scale. For this and other reasons, some evolutionary biologists began to question the assumption that viviparity is always the derived condition and that oviparity is necessarily the ancestral state in all vertebrate groups that include live-bearing species.

BOX 3.1   Dollo's Law and the Supposed Irreversibility of Evolution

In 1893, the Belgian paleontologist Louis Dollo proposed a "law of irreversibility," stating that any complex adaptation, once lost, can never be regained in precisely the same form. In other words, if any complicated biological feature decays during evolution and then disappears (for whatever reason), presumably it can never be precisely recouped. If valid, this law implies a strong asymmetry or bias in the direction of evolutionary loss as opposed to the gain of each complex adaptation. It also implies that any organisms possessing the identical complex adaptation must have inherited that elaborate phenotype from a shared ancestor. Although Dollo's rule makes considerable sense and is a useful generality, all evolutionary laws seem made to be broken, and Dollo's rule is no exception (see Avise [2006] for several empirical examples).

Apart from the empirical violations, there are also theoretical reasons to be suspicious of the universality of Dollo's thesis. For example, Dollo's law presumes that every complex feature has an intricate if not a labyrinthine genetic and ontogenetic basis such that, if the trait were lost, it would not likely reemerge by an identical succession of mutations and selective events. However, biologists know of several instances in which the sudden appearance of a seemingly complicated phenotype (such as the presence of extra legs in a fruit fly or of four wings rather than two) is due to simple mutations in homeotic genes that switch the organism into a different development program (Raff 1996). Thus, although Dollo's law assumes that many genes are modified during the evolution of a complex phenotype, this may not always be true. Furthermore, even if a long and complex chain of genetic causation underlies an elaborate phenotype, one altered link in that developmental chain might cause a complex adaptation to disappear, perhaps merely to reemerge later via a restoration of that crucial link.

Another related point is that some seemingly major interconversions among complex phenotypes might occur rather simply when fine ontoge-

(continued)

BOX 3.1   (*continued*)

netic thresholds are crossed. An example relevant to pregnancy-like phenomena involves the standard textbook distinction between oviparity and ovoviviparity. By definition, an egg-carrying female is oviparous if her fertilized eggs hatch after parturition, but technically she is viviparous if her eggs hatch even a millisecond earlier, just before birth. When a phylogenetic lineage goes back and forth across such narrow definitional lines, even tiny mechanistic steps might cause frequent recurrences of evolutionary conditions that otherwise might seem to be highly complicated and thus implausibly polyphyletic.

To address this issue empirically for reptiles, Lee and Shine (1998) mapped the distributions of live-bearing and egg laying onto a phylogenetic tree for representative extant species. Figure 3.12 depicts a representative subset of this phylogeny, from which it appears that viviparity evolved from oviparity more than a dozen independent times and that few (if any) successful evolutionary transitions occurred in the opposite direction. Thus, although oviparous reptiles might have evolved from viviparous ancestors in some instances, such transitions seem to have been rather uncommon at best. In other words, viviparity appears to have been easier for snakes and lizards to acquire than to jettison during the evolutionary process.

Interestingly, several species of snakes and lizards are polymorphic for oviparity and viviparity, meaning that some populations employ one reproductive mode and other conspecific populations employ the other (Stewart et al. 2010). Surget-Groba et al. (2001) used mtDNA sequences to study the intraspecific phylogeography (Avise 2000) of one such species, *Zootoca* (formerly *Lacerta*) *vivipara* (fig. 3.13) across much of its range in Europe. From these PCM analyses, these authors concluded that a single evolutionary conversion between reproductive modes had occurred (probably in the eastern portion of this species' distribution) and that the direction of change had been from oviparity to viviparity. Results thus proved to be consistent with the general phylogenetic trends described earlier, and they also indicate that such evolutionary transitions can occur quite rapidly (well within a species' geological lifespan, which is probably on the order of one million years or less). One hypothesis for this specific transition is that climatic deterioration during the Pleistocene Ice Ages offered fitness advantages to any gravid females who were able to retain fertilized eggs and actively seek suitable microhabitats where their advanced offspring could be born alive (rather than merely hatched from immobile eggs that had been lain outside).

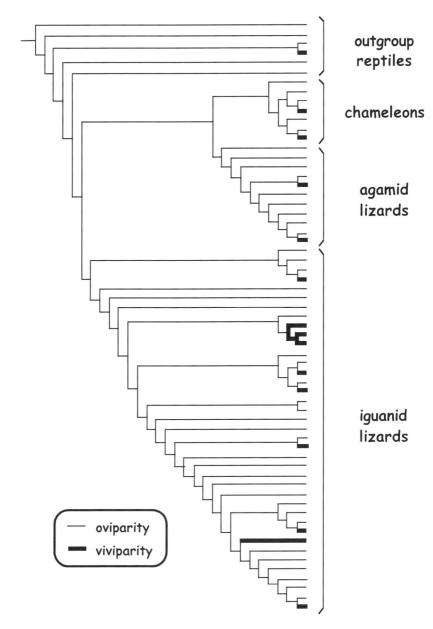

FIGURE 3.12 Phylogenetic tree for more than 60 representative species of lizards and other reptiles, showing more than a dozen independent evolutionary origins of viviparity from oviparity (after Avise 2006, following Lee and Shine 1998).

FIGURE 3.13   The common sand lizard, *Zootoca* (formerly *Lacerta*) *vivipara*, a species that is polymorphic for oviparity and viviparity.

SUMMARY

1. For the many nonviviparous vertebrate species, several other reproductive modalities can be interpreted either as alternatives to pregnancy or as alternative forms of pregnancy, depending upon one's definition of "pregnancy." The traditional antithesis of viviparity is oviparity, in which a gravid female lays eggs rather than births free-living young. However, the varied structures of vertebrate eggs and the diverse hatching times and places in different species can complicate attempts to cleanly demarcate oviparity and viviparity.

2. With regard to the nature of the shed egg, biologically important distinctions relate to the following: (a) fertilized eggs versus unfertilized ova; (b) eggs with relatively impervious casings versus those without such shells; (c) fertilized eggs that hatch before versus after parturition; and (d) eggs that contain substantial food reserves in yolk versus those that do not.

3. All major vertebrate groups include at least some egg-laying species, but only birds are exclusively oviparous. The eggs that birds and reptiles lay are cleidoic or closed, in contrast to the gelatinous eggs of fishes and amphibians, which must be laid in water to prevent desiccation. The cleidoic egg was a key evolutionary invention that facilitated vertebrate colonization of the

terrestrial realm several hundred million years ago, meaning also that the proverbial cleidoic egg came long before the chicken. Even extant mammals include a few egg-laying species: echidnas and the platypus, both of whom are in the order Monotremata.

4. Ovoviviparity is a reproductive mode with elements of both oviparity and viviparity. An ovoviviparous gestation begins with egg-encased embryos that later hatch inside their dam's reproductive tract before eventually being live-born. The relative durations of embryonic gestation within versus outside the egg vary among taxa—from early hatching within the mother to hatching that may occur so late in a pregnancy as to barely precede parturition. Even minor changes in the relative time of hatching can precipitate shifts between egg-laying and live-bearing. Indeed, populations of some species are polymorphic for oviparity and viviparity.

5. To deduce the evolutionary histories of alternative reproductive modes, phylogenetic character mapping has been applied to several vertebrate groups, including fishes and reptiles. In addition to confirming the general fluidity of evolutionary interconversions between egg-laying and live-bearing in various piscine and reptilian taxa, such PCM analyses have indicated that viviparity is often polyphyletic and appears to evolve from oviparity more readily than vice versa.

CHAPTER FOUR

Nonvertebrate Brooders

Although many invertebrate organisms manifest one or another version of viviparity, it is customary to refer to these species as internally producing their offspring rather than as giving birth by standard means of pregnancy. This chapter discusses the diverse expressions of internal (and external) brooding by invertebrate animals. Furthermore, among the external brooders are invertebrate species that retain embryos at one place or another on their bodies, plus other species that encase their embryos in special off-body capsules. Figure 4.1 introduces some of the rationales for including all of these categories of endogenous and exogenous brooding in a book on pregnancy. Pregnancy-like phenomena in nonvertebrates are interesting in their own right, and they also serve as a backdrop for later discussions (in part II) on evolutionary ramifications of pregnancy.

Many of the issues surrounding reproductive modes and pregnancy-like phenomena that arose for vertebrate animals (chapters 2 and 3) often resurface for invertebrates. For example, fertilization may be internal or external, depending on the species, brooding may be absent or present and of variable duration, the sire or the dam (or both) may be the brooding parent, and parental care of offspring can range from nil to extensive. Also, like their vertebrate counterparts, invertebrate embryos require nourishment, which can come from a variety of sources. In marine broadcast spawners with external fertilization, larvae may be either *planktotrophic* (feeding on plankton) or *lecithotrophic* (reliant on more substantial yolk reserves in the egg), and the same is true for postpartum larvae in marine invertebrates that have internal fertilization and brood their zygotes or embryos either internally or externally (Kamel et al. 2010). Understandably, planktotrophic species usually have

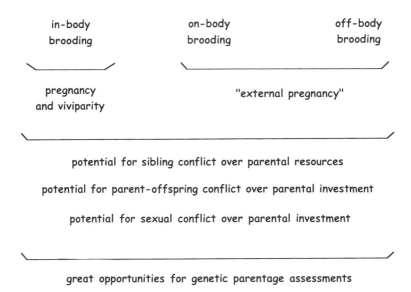

FIGURE 4.1   Three categories of embryonic brooding (in-body, on-body, and off-body) displayed by invertebrate animals. Also shown are how these alternative brooding modes might relate to several biological phenomena typically associated with pregnancy.

higher fecundities (more eggs) than lecithotrophic species because the larger eggs of the latter require more space and more maternal resources (Marshall and Keough 1973; Vance 1973). The "fend-for yourself" feeding lifestyle of planktotrophic larvae places this nutritional tactic at the opposite end of the spectrum from refined matrotrophy, in which embryos receive substantial direct support from their mothers. Overall, the diversity of gestational and feeding modes for brooded invertebrate larvae parallels the varied categories of embryonic gestation and nutrition during vertebrate pregnancies.

The Cast of Players

For invertebrate animals that brood their fertilized eggs and/or embryos (i.e., that display "pregnancy"), brooding can occur in many locations inside or outside the parent's body: in body, on body, or off body. Furthermore, depending on the taxon, the brooding parent may be a female, a hermaphrodite (Avise 2011) or sometimes even a male (Tallamy 2001). To illustrate the taxonomic and biological breadth of brooding arrangements in invertebrate animals, the following sections introduce several phyla, each containing a potpourri of species that sometimes house embryos in specialized or otherwise interesting ways. I begin this survey with annelid worms because these small animals can serve to introduce many evolutionary topics related to different modes of larval brooding and larval feeding.

Annelida

Consider first a rather obscure group of marine worms in the genus *Streblospio* (fig. 4.2). Within and among these tiny species, lecithotrophy and planktotrophy co-occur as life-history alternatives, with various populations showing either or both of these two life-history strategies (Schulze et al. 2000). For example, many females along the Atlantic Coast display "pouch

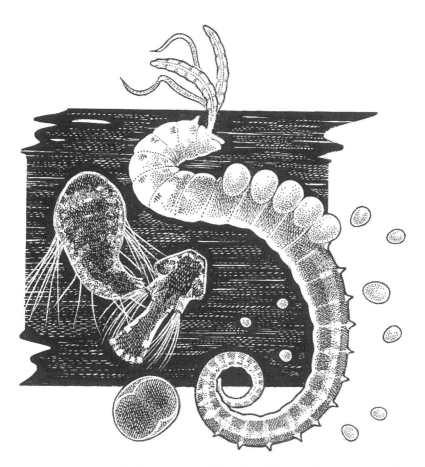

FIGURE 4.2 Benedict's polychaete worm, *Streblospio benedicti*, a poecilogonous species that displays several different forms of larval development.

planktotrophy," in which hundreds of small eggs are stored initially in a dam's dorsal brood pouch, from which plankton-feeding larvae later emerge. In the Gulf of Mexico, most females also release large numbers of planktotrophic larvae, but in this case the small eggs have been stored in their gills ("gill planktotrophy"). Finally, some females in both the Pacific and Atlantic

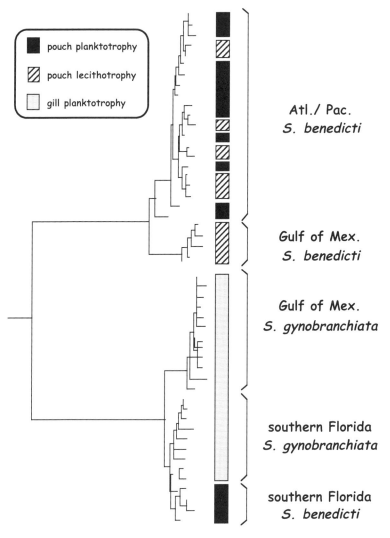

FIGURE 4.3   Molecular phylogeny for populations of polychaete worms in the genus *Streblospio* (after Schulze et al. 2000). Also shown adjacent to the tree are the occurrences of alternative modes of larval development in this taxonomic assemblage.

oceans store large eggs in a dorsal pouch, from which small numbers of yolk-feeding larvae emerge ("pouch lecithotrophy"). Any such intraspecific poly-morphism in larval lifestyle is known as *poecilogony*, a phenomenon that is of special interest to biologists because evolutionary switches between alter-native feeding strategies in effect have been caught in the act in such species.

Using a molecular phylogeny for geographic populations of *Streblospio*, Schulze et al. (2000) plotted the occurrences (and thereby deduced the re-cent evolutionary histories) of contrasting modes of larval developmental (fig. 4.3). This exercise in phylogenetic character mapping yielded two main conclusions. First, as judged by the shallow evolutionary depths within the phylogenetic tree, conversions between larval life histories must occur quite rapidly in these polychaetes. Thus, such life-history features are evolution-arily plastic in at least some invertebrate animals. Second, because experi-mental studies indicated that switches between these larval modes in *Streb-lospio* are not induced by changes in environmental conditions (such as temperature, photoperiod, or food regimen), these alternative developmen-tal trajectories seem to be genetically hardwired. Unfortunately, another key evolutionary question remained unanswered. Because lecithotrophy and planktotrophy (and also pouch brooding versus gill brooding) were closely interspersed within the *Streblospio* phylogeny, Schulze et al. (2000) were un-able to determine from their PCM analyses whether evolutionary switches between the various modes of larval development had been more common in one direction than in another.

The genus *Streblospio* is not alone among polychaete worms in displaying poecilogonous larval development. Another example involves the marine species *Boccardia proboscidea*, which produces both planktotrophic and ben-thic (bottom-dwelling) larvae (Gibson 1997). Indeed, in this case both larval types sometimes emerge from a single-egg capsule of an individual female. Furthermore, some of the offspring within a brood occasionally give a boost to their own development by ingesting "nurse eggs" that their mother also produced and deposited in her egg capsules.

Platyhelminthes

The flatworm *Gyrodactylus elegans* shows a truly bizarre form of pregnancy. In this hermaphroditic species, an egg that has been fertilized internally begins to grow and divide within the mother's uterus. The small assemblage of mitotic cells soon divides unequally, generating a second embryo that starts to develop inside the first, but not until the first daughter embryo is released from the parent. A third embryo then begins to develop within the second, a fourth within the third, and so on, eventually yielding as many as 2,500 genetically identical progeny in four weeks (Baer and Euzet 1961, as cited in Craig et al. 1997). This Russian-doll configuration of clonemates may be nearly unique to these fish-parasitic flatworms, but it illustrates the lengths to which evolution can go in generating offspring via pregnancy.

Cnidaria

In most corals and other cnidarians, both sperm and ova are broadcast into the ocean, such that fertilization is external, and the resulting larvae develop in the plankton without benefit of direct parental care (Veron 2000). However, in a substantial minority of species, ova are retained either by the female or by a hermaphrodite, such that fertilization is internal, and the planula larvae begin life within the body of a parent. In a survey of reproductive modes for nearly 200 coral species from several of the world's major reef areas (the Caribbean, tropical Pacific, and Red Sea), Richmond and Hunter (1990) tallied the number of species that qualify as broadcast spawners without brooding (168 species), hermaphroditic brooders (11 species), and gonochoristic (separate-sex) brooders (7 species). Even closely related coral species sometimes show different modes of larval development. For example, in the genus *Acropora* most species, including *cervicornis* (fig. 4.4), are broadcast spawners, but several others, such as *palifera* and *striata*, produce their progeny internally (Ayre and Miller 2006).

Most coral species are hermaphroditic (Harrison and Wallace 1991), and in some species the dual-sex colonies occasionally self-fertilize, an incestuous behavior that nonetheless may be advantageous when mates are rare or sperm are in limited supply. Hermaphroditic species that retain (rather than broadcast) their ova may be predisposed to brood their young, so perhaps it is not too surprising that brooding and selfing co-occur in several coral species (Richmond and Hunter 1990; Sherman 2008). Another route to "pregnancy" in corals occurs in several taxa in which the brooded larvae are produced asexually (Stoddart 1983; Ward 1992).

A relatively small number of sea anemones, such as *Epiactis prolifera* (fig. 4.5) also brood their larvae. In this species, each individual begins life as a female but later becomes a hermaphrodite (a functionally dual-sex specimen). Eventually, the hermaphrodite will brood selfed or outcrossed progeny externally on the body's central column (Bucklin et al. 1984).

Factoid: Did you know? Like many other invertebrates (and plants), most corals can propagate not only sexually but also asexually via "cuttings." Such clonal reproduction often occurs when a branch of coral breaks off from the parent colony and falls to the substrate, where it initiates a new colony.

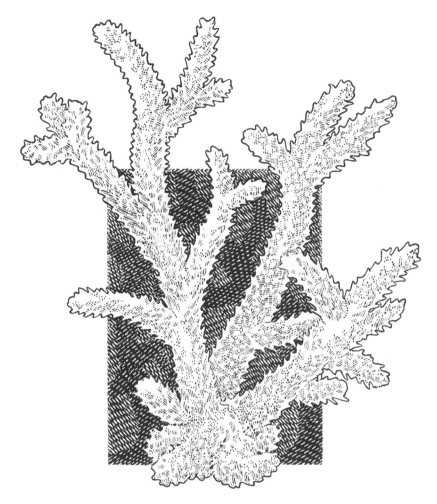

FIGURE 4.4   The Caribbean staghorn coral, *Acropora cervicornis*. Although this and most other staghorn corals are broadcast spawners, a few species in the genus brood their larvae.

Factoid: Did you know? In *Epiactis prolifera*, eggs are fertilized in the adult's digestive cavity, after which the motile larvae swim out of the parent's mouth and install themselves on the adult's outer body, where they incubate and grow into autonomous little anemones.

FIGURE 4.5   The brooding sea anemone, *Epiactis prolifera*, a cnidarian species with a highly peculiar mode of fertilization and external brooding (see the factoid).

Bryozoa

Bryozoans (phyla Entoprocta and Endoprocta) are tiny solitary or colonial marine creatures that superficially resemble corals and consist of specialized functional units known as zooids. They are of special interest in the context of pregnancy because many species possess placenta-like analogues via which larvae receive maternal nutrients during their development either within or outside the parent's body cavity (Dyrynda and Ryland 1982). Depending on the particular Bryozoan group, placental contact between mother and larva can sometimes be intimate, even to the point of fusion and substantial nutrient transfer between fetal and parental tissues. Comparative analyses and phylogenetic character mapping have demonstrated that various categories

> **Factoid: Did you know?** Bryozoans are often referred to as moss animals, sea mosses, or sea mats because their branched or matted colonies can resemble seaweeds when they spread plantlike across rocks or other marine substrates.

of placentation have arisen multiple times in Bryozoa and thus are polyphyletic (Ostrovsky et al. 2009).

Echinodermata

Although brooding is relatively unusual in this group of marine animals, all major echinoderm taxa include at least some species with larval brooding (Hart et al. 1997). For example, in some sea stars (class Asteroidea) and brittle stars (Ophiuroidea), a female extrudes yolk-rich eggs and then forms a brooding chamber by bending her arms ventrally around her central disk. In *Leptasterias hexactis* (fig. 4.6), brooding therein proceeds for about 40 days postfertilization before the juveniles develop tube feet and walk away.

> **Factoid: Did you know?** In some hermaphroditic sea stars that brood their young internally, juveniles routinely eat some of their siblings while still housed jointly in their pregnant parent's gonadal tissue (Byrne 1996).

FIGURE 4.6   The sea star, *Leptasterias hexactis*, an echinoderm species that broods its eggs and larvae.

> **Factoid: Did you know?** In the brooding sea cucumber, *Leptosynapta clarki*, juveniles remain in their pregnant mother's ovary for about 6 months before exiting by rupturing her body wall (Sewell 1994).

Other echinoderms show different forms of brooding. Although most sea cucumbers (Holothuroidea) are broadcast spawners with planktonic larvae, more than 40 holothurian species brood their young in various locations such as under their sole, in tentacles, in brood pouches, or, internally, in either the ovary or the body cavity (Smiley et al. 1991). In the sea cucumber, *Synaptula hydriformis*, fertilized eggs hatch, and larvae develop inside a perivisceral coelom within the parent's body before being live-born. This species is of interest for two additional reasons: it is a self-fertilizing hermaphrodite, and it is matrotrophic, meaning that the gestating young receive at least some parental nutrition in excess of that supplied by the yolk (Frick 1998). Brooding of one form or another also occurs in various other echinoderms, including some sea urchins and sand dollars (Echinoidea), where the phenomenon has apparently evolved independently on at least a dozen occasions (Emlet 1990).

Mollusca

All major groups of mollusks, including snails (class Gastropoda), bivalves (Bivalvia), chitons (Polyplacophora), and squids and their allies (Cephalopoda), encompass at least some species that brood their larvae. Fertilization in such species is typically internal, either following mating or by some direct means of sperm uptake from the environment.

In gastropods, brooding either in the mantle cavity or within an enlarged oviduct is quite common, as well as polyphyletic (Köhler et al. 2004). Several examples from the marine realm involve *Littorina* periwinkles in the family Littorinidae (Johannesson 1988; Reid 1990), some *Planaxis* snails (Planaxidae), and some slipper limpets (Calyptraeidae), including *Crepidula fornicata* (fig. 4.7). Individuals in the last species are sequential hermaphrodites (Avise

> **Factoid: Did you know?** As a part of the mating process, many snails spear each other with calcareous "love darts" containing a mucus that decreases sperm digestion in the recipient and thereby increases the chance of fertilization by the successful dart thrower.

FIGURE 4.7   Stack of slipper limpets, *Crepidula fornicata*, a mollusk that broods larvae under its shell.

Factoid: Did you know? In slipper limpets an individual typically begins life as a male but later becomes a female, in a reproductive system known as *protandry*. In other animal species showing sequential hermaphroditism, an individual may begin life as a female but later becomes a male, in a reproductive system known as *protogyny*.

2011) that live in stacks in which the bottom individual is a female and higher individuals are younger males with whom she mates. The mother then broods offspring underneath her caplike shell, packaged in dozens of capsules, each containing dozens to hundreds of embryos (Proestou et al. 2008).

Somewhat like the *Streblospio* worms and the *Acropora* corals described earlier, periwinkles and slipper limpets provide examples of poecilogony because even closely related populations or species in each of these genera sometimes show different strategies of larval development, including planktotrophy and brooding (Hoagland 1986; Collin 2004). Thus, these species again demonstrate that larval ontogeny can be labile during marine invertebrate evolution. The genus *Crepidula* is also of special interest because well-preserved fossils, complete with brooded larvae, have been discovered in some of its species (Herbert and Portell 2004).

An elegant form of off-body larval brooding occurs in many whelk species (Buccinidae and Melongenidae) such as *Busycon carica* (fig. 4.8), in which a

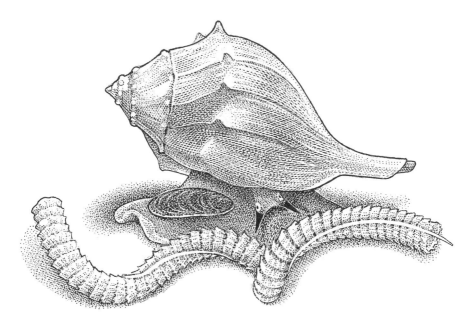

FIGURE 4.8   A female knobbed whelk (*Busycon carica*) and her string of egg cases.

Factoid: Did you know? In some marine snails such as the Scotch Bonnet (*Phalium granulatum*), a female extrudes a cylindrical tower of egg cases, upon which she sits like a queen on her throne.

---

previously mated female lays a beautiful string of about 80 leathery capsules, each of which houses several dozen embryos (Walker et al. 2005, 2007; Avise et al. 2004). Many other marine gastropods, including wentletraps (Epitoniidae), turritellas (Turritellidae), trivias (Eratoidea), moon snails (Naticidae), bonnets (Cassidae), dove shells (Columbellidae), tulip conchs (Fasciolariidae), and cowries (Cypraeidae), similarly lay benthic egg capsules of diverse and often lovely structural designs.

The terrestrial snail *Cantareus (formerly Helix) aspersus* (fig. 4.9) illustrates another form of off-body brooding (Herzberg and Herzberg 1962). In this hermaphroditic species, copulation is sometimes reciprocal, with each individual simultaneously inseminating its partner. Several days later, each "pregnant" parent digs a 1-inch nest in moist soil, wherein it deposits about 80 fertilized eggs. About two weeks later, the embryos hatch and crawl away.

FIGURE 4.9   The brown garden snail, *Cantareus aspersus*, a nest-brooding species.

Internal brooding in bivalve mollusks is taxonomically widespread and can take many forms (Hain and Arnaud 1992). In such species, internal fertilization typically occurs when a gravid female or a gravid hermaphrodite "inhales" sperm through a siphon that it also uses to filter water for food and oxygen. Depending on the species, the resulting zygotes and embryos may be released quickly or perhaps retained either in the body cavity or gill chambers for varying lengths of time. In the marine species *Adacnarca nitens*, each female broods 50–70 offspring internally, in clusters of embryos housed within her branchial chamber (Higgs et al. 2009). In freshwater mussels in the family Unionidae (fig. 4.10), an adult may brood thousands to millions of larvae in its gills, where the "glochidial embryos" are initially enclosed in a vitelline membrane that provides intimate physical contact with the parent and is suspected to play a nutritive role early in embryonic development (Schwartz and Dimock 2001). After this gestational phase, the discharged larvae typically attach to a host fish, where they continue developing and eventually metamorphose into juvenile mussels (Rogers-Lowery and Dimock 2006).

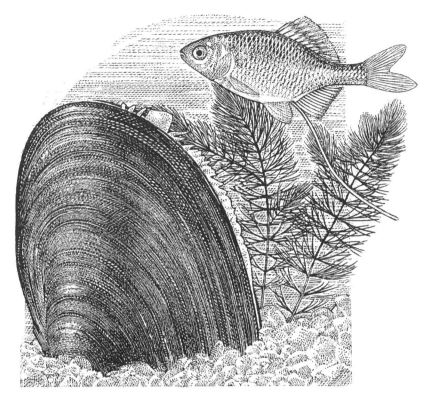

FIGURE 4.10   Freshwater unionid mussel and bitterling fish. These mussel species brood their larvae internally until the latter transfer to a secondary host (a fish) to continue their development.

Factoid: Did you know? Some unionid mussels have evolved special wormlike protuberances that they actively wave to attract a host fish, who is then colonized by the bivalve's shed larvae.

In general, larval brooding in bivalves and other benthic (bottom-dwelling) invertebrates is presumed to be a useful reproductive tactic especially in harsh environments, where embryonic survival might otherwise be extremely low (Poulin et al. 2002; Galley et al. 2005; Heilmayer et al. 2008). One potential disadvantage of brooding (in contrast to planktotrophy) is that the relatively advanced larvae born of a "pregnant parent" may have rather limited opportunities for dispersal (Carlon 2002). However, in some cases even brooding species can be effective colonizers (Johannesson 1988), especially when they

FIGURE 4.11   Egg-brooding female of the giant deep-sea squid, *Gonatus onyx*.

take advantage of some other dispersal mechanism such as floating (High-smith 1985), "kelp rafting" (Helmuth et al. 1994), or hitching rides on second-ary hosts (as in the freshwater mussels).

A few mollusks in the class Cephalopoda also brood their young. In squids and octopuses, internal fertilization follows mating in which the male trans-fers a spermatophore to his mate's mantle cavity. Although biologists used to think that each female then deposits her fertilized eggs on the sea floor and leaves them to develop on their own, recent observations reveal that individuals in some squid species hold such eggs in their arms (attached to small hooks) until the embryos hatch (Seibel et al. 2005). In the giant deep-sea squid, *Gonatus onyx* (fig. 4.11), females have been photographed while arm-carrying thousands of egg-encased embryos in huge clusters. A female octopus also guards and tends her fertilized eggs, which she typically lays by the thousands in strings that she hangs from the ceiling of her lair.

Arthropoda

Many species in this largest invertebrate phylum brood their young in one fashion or another. For example, females in many crustacean species (subphy-lum Crustacea) carry embryos in special pouches inside or outside their bodies. In the suitably named "opossum shrimps" (Mysida) and in many spe-cies of isopods (Isopoda) and amphipods (Amphipoda), that pouch is called

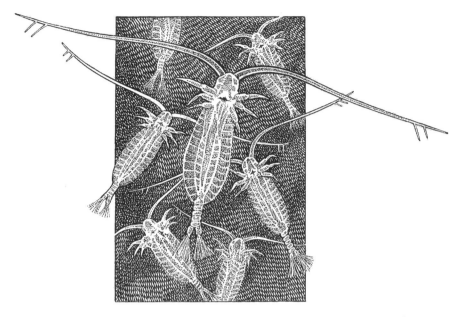

FIGURE 4.12   Copepods (Copepoda) are among the many arthropod groups in which various species brood their young.

a *marsupium*, and it is located on the female's thorax. Many other crustaceans, including fairy shrimps (Anostraca), water fleas (Cladocera), and copepods (fig. 4.12), also brood their eggs in special dorsal, ventral, or lateral sacs before releasing them to the outside world (Ohman and Townsend 1998; Logerwell and Ohman 1999). Even many sessile barnacles (Cirripedia) manage to mate and brood internally fertilized eggs (Crisp 1959; O'Riordan et al. 1992). In these typically hermaphroditic species, an individual copulates with close neighbors using an extendable penis, after which the eggs are brooded and later released as nauplii larvae, which may spend weeks in the plankton.

In many species of crabs, shrimps, crayfish, and lobsters (Decapoda), a gravid female typically cups her abdomen under her cephalothorax and thereby forms a spawning chamber into which she lays and then fertilizes her eggs, typically using dissolved spermatophore packages that she had received from earlier matings with one or more males (Walker et al. 2002; Toonen 2004; Gosselin et al. 2005; Yue et al. 2010). She then carries the eggs and juveniles (fig. 4.13) on her *pleopods* (swimming legs) for as long as several months, all the while fanning and protecting the hatchlings until they are mature enough to depart and fend for themselves.

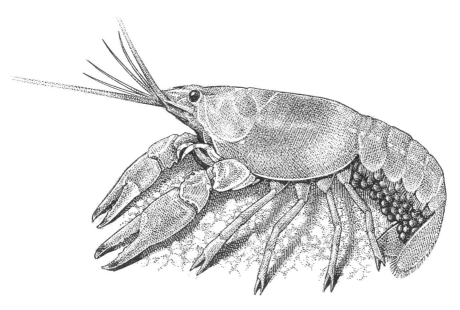

FIGURE 4.13   Mother and brood in a freshwater crayfish, *Orconectes placidus.*

Factoid: Did you know? A lobster mother may carry as many as 30,000 embryos.

In several species of land crabs from Jamaica, females show perhaps even more parental attention by actively tending their broods in off-body sites. In the bromeliad crab, *Metopaulias depressus* (fig. 4.14), each mother raises her young in the base of a water-filled bromeliad leaf. There, she circulates the water, removes detritus, feeds her young, and protects them from predators. She even hauls in empty snail shells that provide calcium to her offspring and serve as a pH buffer. In another Jamaican endemic (*Sesarma jarvisi*), the adult crab actively delivers water to an empty snail shell or turns the adopted shell upside down to collect rainwater. For several months, the snail shell then becomes a relatively safe nursery for the crab's embryos.

Factoid: Did you know? Although Jamaica's bromeliad crab is fully nonmarine, many other so-called land crabs around the world liberate their hatched larvae into the sea, where they continue their development.

FIGURE 4.14   The bromeliad crab, *Metopaulias depressus*, a terrestrial crustacean in which mothers tend their brooded young in special off-body incubators.

Insects (Insecta) are the most species rich of extant arthropods, and although most insects lay eggs (i.e., are oviparous), females in some taxa retain fertilized eggs within the uterus and later give birth to live young (Meier et al. 1999; Agarwala and Bhadra 2010). Thus, such species are ovoviviparous. A few insects even show *facultative ovoviviparity*, wherein a female either lays

> **Factoid: Did you know?** Researchers have managed to promote the evolution of ovoviviparity in experimental fruit-fly populations by selecting for females with increased live-bearing behavior.

fertilized eggs or delivers already-hatched young, depending on the environmental circumstances (Markow et al. 2008).

In a select few invertebrates, males (rather than females) tend the broods (Clutton-Brock 1991; Tallamy 2001). Among the arthropods, the two most famous examples are giant water bugs in the insect family Belostomatidae (Smith 1976, 1997) and the 1,200-plus sea spider species in the subphylum Pycnogonida (King 1973; Shuster and Wade 2003; Bain and Govedich 2004). In the belostomatids, a male water bug typically carries 30–160 fertilized eggs that his mate has cemented to his back (fig. 4.15).

In the pycnogonid sea spiders (fig. 4.16), each male typically totes dozens to hundreds of fertilized eggs in masses glued either to his ventral surface or to a pair of specialized legs known as *ovigers*. In the sea spiders and giant

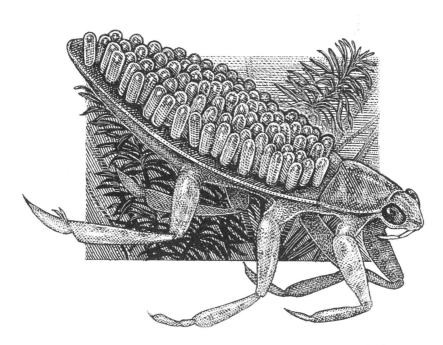

FIGURE 4.15   A male giant water bug carrying fertilized eggs cemented to his back.

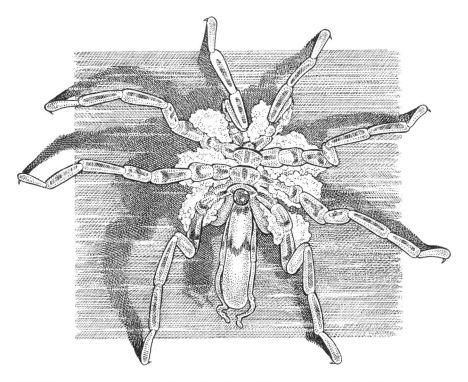

FIGURE 4.16   A male sea spider (*Ammothea hilgendorfi*) carrying masses of fertilized eggs on his legs.

water bugs alike, males may retain the developing eggs and embryos for several weeks or even months (Smith 1974; Tomaschko et al. 1997).

One prediction of selection theory is that a male should offer substantial care of this sort only if the members of his brood carry copies of his own genes (Trivers 1972; Thornhill 1976). For water bugs, observations suggest that each "pregnant" male does indeed have high assurance of paternity for his brooded young (Smith 1979a, 1979b), in part because a male copulates repeatedly with a female before receiving her eggs; in addition, for sea spiders, genetic-parentage analyses have confirmed that each brood-carrying male is indeed the biological sire of the embryos that he tends (Barreto and Avise 2008, 2010, 2011).

Factoid: Did you know? Although paternal care of offspring is relatively rare in invertebrates, such behavior has evolved independently on at least 17 occasions in these animals (Tallamy 2001).

Urochordata

Many of the 1,000-plus species in this subphylum of Chordata consist of hermaphroditic individuals that are male and female simultaneously. Sperm are shed into the sea, whereas eggs typically retained by an adult are fertilized internally by male gametes sucked into a hermaphrodite's body through special incurrent siphons. The resulting larvae then brood internally before being released live by the viviparous parent.

Factoid: Did you know? Urochordates are also known as tunicates, or sea squirts, the latter name deriving from the fact that these small, sessile, baglike animals often squirt seawater when stepped on or otherwise disturbed.

Plants

A botanical analogue of pregnancy also exists (Farnsworth 2000). *Vivipary* is the precocious development of sexual progeny directly on the parent plant (unlike the germination of a seed elsewhere, such as in the soil). True vivipary entails the germination of sexual seeds that for a time remain attached to and develop directly on the dam. In other words, viviparous seeds typically germinate and grow within the fruit prior to abscission from the maternal parent. This phenomenon is rare in plants, having been described in only about 50 species (Elmqvist and Cox 1996). Examples are mangrove trees in the genus *Rhizophora* (Farnsworth and Farrant 1998); alpine bluegrass, *Poa alpina* (fig. 4.17); spreading schiedea, *Schiedea diffusa* (Wagner et al. 2005), which is endemic to Hawaii; and several coastal cacti (fig. 4.18) native to northwestern Mexico (Cota-Sánchez et al. 2007). Another type of "live-bearing" in plants, sometimes called *pseudovivipary*, transpires when asexual (rather than sexual) propagules grow on the parent (Beetle 1980; Elmqvist and Cox 1996).

One could also argue that all seed plants are in effect viviparous in the sense that during seed development an embryo is brooded and nurtured within an ovule, initially while attached to the maternal plant. The seed coat (*testa*) of the mature seed is in fact maternal tissue that encases the gestating conceptus. From this biological perspective, it is perhaps unfortunate that vivipary has acquired a much more restricted meaning in botany.

Evolutionary Transitions

As was true for vertebrate pregnancies (chapter 2), standard prerequisites for larval brooding by invertebrates are the retention of ova by females and

FIGURE 4.17   The alpine bluegrass, *Poa alpina*, a "pregnant" plant species in which sexually produced seeds germinate while still attached to the mother.

internal fertilization. This means that during any evolutionary transition from nonbrooding ancestors to internal-brooding descendents, one of the parents (typically the dam) assumes added responsibility for offspring care. How this comes about in particular instances undoubtedly varies across taxa. Here are merely a few of the evolutionary considerations and some empirical examples.

Planktotrophy to Lecithotrophy

Conventional wisdom for marine invertebrates held that planktotrophy was the ancestral lifestyle from which lecithotrophy recurrently evolved in various taxonomic groups. For example, lecithotrophy probably evolved from

FIGURE 4.18   The Mexican columnar cactus, *Pachycereus pringlei*. Some species in this genus have viviparous fruits containing sexual progeny that continue to grow directly on the parent plant.

planktotrophy at least four independent times in asterinid starfish (Hart et al. 1997). This evolutionary directionality can be rationalized by supposing that whenever complex feeding structures are lost by larvae in a planktotrophic lineage, such adaptations may be difficult to regain. On the other hand, several probable instances have been identified (e.g., in littorinid and calyptraeid snails) in which larval feeding (planktotrophy) appears to have reevolved in nonfeeding (lecithotrophic) lineages (Reid 1990; Collin 2004). Overall, studies on species representing several invertebrate phyla, including Mollusca and Echinodermata, have revealed (as in the *Streblospio* polychaetes and *Acropora* corals described earlier) that modes of larval development in marine invertebrates can switch back and forth quite rapidly in geological time and thus are not always tightly constrained in evolution. Compared to plankto-trophy, lecithotrophy often registers an increased investment by the mother in her progeny, albeit indirectly via a richer yolk that she has provisioned to each egg. However, brooding (and, in the extreme, matrotrophy) can take parental investment to even greater heights.

Onward to Female Brooding

Freshwater jawless leeches (family Glossiphoniidae, phylum Annelida) are remarkable not only for their bloodsucking behavior but also for their refined parental care of offspring. All glossiphoniids use their flattened bodies to cover their developing egg capsules, but some species, such as *Marsupiob-della africana* (fig. 4.19), go at least one step further by carrying their eggs and young in a special internal brood pouch.

Sawyer (1971) used evolutionary logic in conjunction with a suspected phylogeny for these leeches to deduce the probable history of this transition to internal pregnancy. In Sawyer's reconstruction (fig. 4.20), the process be-gan in an ancestor that laid unattended eggs from which larvae hatched and simply crawled away. Then, in stage II of the evolutionary progression, the

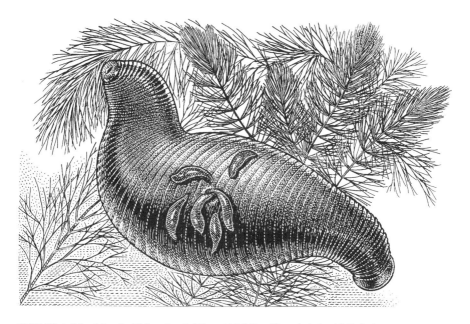

FIGURE 4.19   A South African leech (*Marsupiobdella africana*) giving birth to larvae. This species broods its young internally.

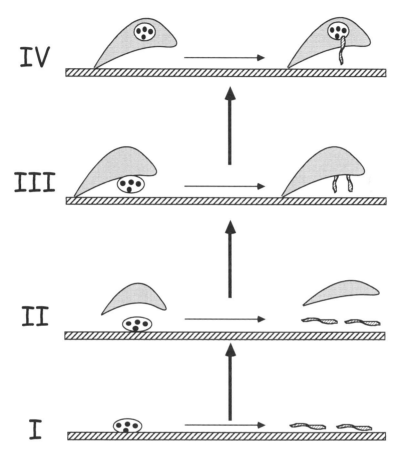

FIGURE 4.20   Four presumptive stages (see the text) in the evolutionary transition from egg laying to live-bearing in the leech family Glossiphoniidae (after Sawyer 1971).

parent presumably tended the substrate-attached eggs by covering them with its body. In stage III, the egg capsules became cemented to the parent's ventral surface. Finally, in evolutionary step IV, the parent's body became reshaped to curve around the egg capsules and thereby form an internal pouch, within which larvae later hatch and are "live-born." In the Annelid subclass Hirudinea as a whole, extant species are known that represent all of these presumed stages in the evolutionary progression from oviparity to viviparity.

Similar kinds of evolutionary scenarios can be envisioned for various other viviparous marine invertebrates. In gastropod mollusks, for example, in many oviparous species a female simply lays unattended egg capsules, but each female cowry (Cypraeidae) sits on her clump of eggs for many days until they hatch, and females in several viviparous groups retain their egg capsules internally until they hatch and the young are live-born.

Evolutionary reconstructions can be similarly tailored for some viviparous invertebrates that inhabit the terrestrial realm. For example, most *Drosophila* fruit flies are oviparous, but developmental profiles vary greatly within and among species, with females in a few species, including *sechellia* and *yakuba*, sometimes retaining fertilized eggs within their uteri before giving birth to live young (Markow et al. 2008). Such ovoviviparity might be adaptively advantageous under particular environmental circumstances, such as when oviposition sites are scarce or when fertilized eggs might need the added protection that internal brooding can provide. More generally, as is also true for various vertebrate taxa (chapter 3), ovoviviparity in invertebrates might be a rather common intermediate step during evolutionary interconversions between oviparity and viviparity (Sellier 1955; Markow et al. 2008). Indeed, examples of viviparity in insects have long been known (Hagan 1948), and Meier et al. (1999) suggest that viviparity (including ovoviviparity) has evolved from oviparity more than 60 times within the dipteran flies alone. Minimally, all that need be entailed in such transitions is the retention of fertilized eggs that hatch within the female prior to (rather than after) oviposition.

Onward to Paternal Care of Offspring

In a small minority of invertebrate and vertebrate animal species, males take the prime responsibility for offspring brooding. How such paternal investment might have evolved has been debated extensively (Goodwin et al. 1998; Dulvy and Reynolds 1997; Andersson 2005; Reynolds et al. 2002), with evolutionary scenarios often differing depending on the taxa (such as birds versus fishes) being considered (Reynolds et al. 2002). (Some of the evolutionary scenarios for fishes are presented in chapter 7.)

Larval brooding by invertebrate males is also an uncommon phenomenon; it is displayed by only a few groups, including water bugs (Belostomatidae) and sea spiders (Pycnogonidae). For the water bugs, a combination of field observations and phylogenetic (PCM) analyses led Smith (1997) to create the following scenario for the evolution of the male's peculiar habit of carrying fertilized eggs on his back. In many nonbrooding belostomatid species, females typically lay their zygotes on emergent aquatic vegetation, where the embryos develop and respire using atmospheric oxygen that diffuses to them. According to Smith (1997), some gravid females may have accidentally laid their eggs on mate-guarding males instead. Furthermore, such males may not have been unduly inconvenienced by this burden because any egg-toting costs (such as impaired swimming or visibility to predators) might have been outweighed by an increase in the male's genetic fitness through increased progeny survival (Kraus et al. 1989). After males have been "captured" in such a paternal-care tactic, females would probably come under increased selection pressure to put energy into producing more eggs and eventually surrendering all postzygotic care to their mates (Tallamy 1995). The water bugs' habit of alternating copulation with oviposition probably gives each male a high assurance of paternity for the brood he tends, so this particular mating behavior, too, likely arose during this evolutionary transition to exclusive paternal care of offspring.

Conflicts Promoted by Brooding

For invertebrate animals, brooding almost invariably entails packaging many full-sib or half-sib embryos in cramped quarters within, on, or adjacent to the body of one parent (usually the dam). Such communal housing in conjunction with the particular genetic relationships among broodmates sets the evolutionary stage for various conflicts of interest over finite resources (Parker et al. 2002). The following sections introduce these topics, which we then explore further in part II.

Sibling Conflict

Sibling rivalries (Mock and Parker 1997) are likely to be exacerbated in vertebrate pregnancies and invertebrate broods alike because the offspring are in close contact and have ample opportunities to interact. The intensity of such conflicts can vary according to many factors, such as the size of the brood relative to the availability of resources (e.g., space, oxygen, nutrients), and to the broodmates' genetic relationships, which themselves are a function of the reproductive mode (e.g., sexual vs. asexual) and the mating system (e.g., monogamy vs. polygamy) of the parent(s). The resolution of such intrabrood conflicts may be reflected in outcomes as subtle as differential growth rates among the broodmates or as overt as sibling cannibalism (see box 2.1).

Parent-Offspring Conflict

From their own selfish genetic perspectives, a parent and its offspring typically have different "ideas" about how best to invest finite parental resources (Trivers, 1974). Typically, from a parent's perspective the optimal strategy is to supply less than what an offspring would ideally desire. This fact yields an inherent intergenerational conflict of interests among family members, the magnitude of which depends on many additional factors, such as the mating system (which can influence genetic relationships within each family), the prospects for future reproduction by the parent, and the relative contributions of dam and sire(s) to the rearing of progeny (Parker 1985). An extensive body of evolutionary theory discusses how parent-offspring conflicts might be resolved in particular biological circumstances (Trivers 1972; Maynard Smith 1977; Parker and Macnair 1978; Macnair and Parker 1978, 1979; Charnov 1982; Mock and Forbes 1992; Godfray 1994; Lessels and Parker 1999; Parker et al. 2002). In general, vertebrate pregnancy and invertebrate brooding both tend to heighten the intergenerational tension because in such circumstances a parent and its offspring interact directly and often intimately (Zeh and Zeh 2000; Crespi and Semeniuk 2004).

Parent-Parent Conflict

From their own selfish genetic perspectives, males and females in sexual species typically have very different approaches to how best to optimize personal fitness (see chapter 7) despite the fact that they must somehow collaborate to produce successful offspring. This sexual conflict of interests routinely leads to sexually antagonistic selection that can operate at any stage (or all stages) of the reproductive process, ranging from prezygotic mate choice and courtship behaviors, to fertilization modes, to tactics of postzygotic parental care (Arnqvist and Rowe 2005). In general, the phenomena of pregnancy (in vertebrates) and larval brooding (in invertebrates) tend to heighten the intersexual tensions because huge disparities then characterize the mode and magnitude of personal investment in progeny by sires and dams. Thus, what is best for the genetic fitness of the goose is not necessarily what is best for the gander.

Genetic Parentage Assessments

All forms of invertebrate brooding in which the female parent remains associated with the brood (and can be collected with it) offer researchers wonderful opportunities to determine who sired the embryos and thereby to deduce the genetic mating system of a species. Under the "one-parent-known" statistical model (Selvin 1980), the process works generally as follows (see the appendix). Using suitable molecular techniques in a genetics laboratory, a researcher assays the female brooder and her known biological offspring at multiple marker genes (loci). At each locus, each diploid embryo inherited one allele from its known dam so the other allele that it is observed to carry must have come from its (otherwise unknown) sire. The researcher then accumulates such genetic data for several polymorphic marker loci and for multiple embryos within the brood and thereby deduces the multilocus genotype of each brood's sire(s). By repeating this protocol for many pregnancies or broods, a researcher can unveil the genetic mating system of a population or species. Sometimes it is possible to deduce not only the number of sires and their relative contributions to each brood but also to map the spatial

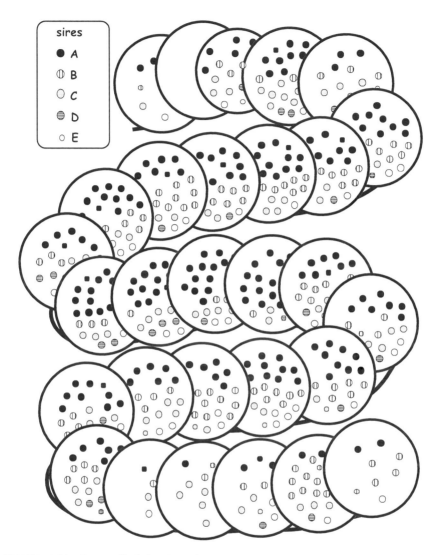

FIGURE 4.21   Genetically deduced sires for several hundred embryos along 29 successive egg capsules in an egg-case string from a female knobbed whelk (after Walker et al. 2007).

positions of their embryos within a female's egg case or her brood pouch. Examples involving the egg capsules of a marine whelk (*Busycon carica*) and the brood chambers of a freshwater crayfish (*Orconectes placidus*) are diagrammed respectively in figure 4.21 and figure 4.22.

An analogous but gender-reversed process is employed to deduce maternity and the mating system for species in which the brooding parent is the

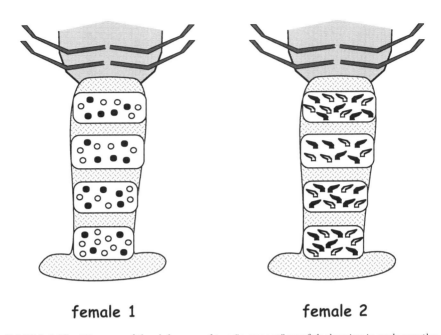

female 1　　　　　　　　female 2

FIGURE 4.22　Diagrams of the abdomens of two "pregnant" crayfish showing in each case the spatial arrangements of progeny from two genetically deduced sires (open versus closed symbols). Left, unhatched eggs in the brood of one female; right, hatchlings in the brood of a second female (after Walker et al. 2002).

sire rather than the dam. An empirical example involving the sea spider, *Ammothea hilgendorfi*, is presented in figure 4.23.

In principle, similar kinds of genetic analyses can be conducted on non-brooding species in which neither biological parent is known at the outset, but the detective process is far easier when one known parent can be collected together with its brooded young.

SUMMARY

1. Internal brooding by viviparous invertebrates is the evolutionary analogue of pregnancy in vertebrate animals. Many invertebrate phyla ranging from Annelida (polychaetes and other worms) to Cnidaria (corals) to Echinodermata (sea stars and their allies) to Mollusca (snails and their allies) to Arthropoda (insects and their allies) include at least some species that give birth to live offspring after having brooded their larvae internally. Indeed, even a few plant species are in effect viviparous.

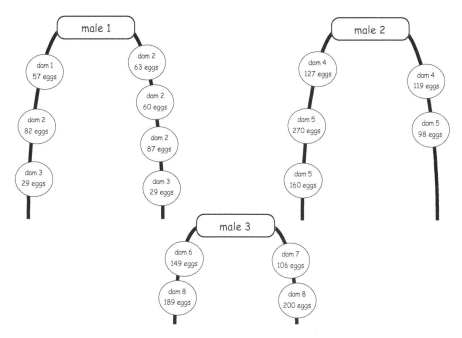

FIGURE 4.23   Diagrammatic representation of broods carried on the specialized legs (heavy lines) of three male sea spiders (after Barreto and Avise 2010). Each circle represents an egg cluster composed of the indicated numbers of eggs (embryos). Genetic analyses of the sires and each of their progeny were used to decipher the dams of these embryos. Note that all of the embryos within a cluster proved to have the same mother but that each male had received egg clusters from at least two females. Overall, this parentage study involving nearly 1,400 embryos from 13 "pregnant" males indicated that this population of *Ammothea hilgendorfi* has a polygynandrous mating system because members of both sexes proved to have had multiple sexual partners during each breeding episode.

2. In addition to the internal brooders, females (and, far more rarely, males) in various other invertebrate taxa carry their developing young on the outside of their body or brood their larvae in special off-body nests or capsules. Altogether, the wide diversity of brooding and feeding modes for invertebrate larvae parallels the great diversity of gestational phenomena for vertebrate embryos.

3. For several invertebrate taxa, phylogenetic character mapping (PCM) in conjunction with other lines of evidence has been used to reconstruct the probable course of evolutionary interconversions among alternative reproductive modes. Some such analyses have even been conducted across microevolutionary timescales because several invertebrate taxa display the

phenomenon of peociliogony, wherein alternative modes of larval brooding and feeding coexist as polymorphisms within particular species.

4. Some of the evolutionary transitions provisionally reconstructed for various invertebrate groups are as follows: (a) from ancestral planktotrophy (plankton feeding) to lecithotrophy (yolk feeding) by larvae; (b) from off-body larval brooding to on-body larval brooding to within-body brooding by dams; and (c) from off-body larval gestation without parental care to on-body brooding by sires. The latter transition (to a tactic of high paternal investment) has been rare and apparently has been accompanied by a high assurance of genetic paternity for the brooding males.

5. For sexually reproducing species, invertebrate brooding (like vertebrate pregnancy) tends to exacerbate several inherent evolutionary conflicts of interest: between siblings within a brood, who compete for finite resources; between the two parents, who have different optimal tactics for enhancing personal genetic fitness; and between brooding parents and their progeny, who have different perspectives on the optimal magnitude and distribution of parental investment in offspring.

6. Another important feature of invertebrate brooding that is analogous to that of vertebrate pregnancy relates to genetic parentage analyses. Because full-sib and half-sib embryos are associated physically with their brooding parent (typically the dam), molecular markers can be employed to deduce the sire(s) of each brood and thereby reveal the genetic mating system of a population. For species in which the brooding sex is the male, analogous determinations of genetic maternity and mating systems can likewise be accomplished. In some cases the arrangements of different full-sibships within a brood can even be microspatially mapped inside each brooding chamber.

CHAPTER FIVE

Human Pregnancy in Mythology and in Real Life

Each human pregnancy technically begins when a sperm cell and an ovum unite inside a woman and produce a fertilized egg or zygote, whose nucleus contains 2 nearly matched sets of genetic material, one from the father and the other from the mother. This never-before-seen mixture of genes interacts to help direct the progeny's biological life—from preembryo to the person's death perhaps decades later. However, in the first 2–4 rounds of cell division, most of the RNA and protein molecules that orchestrate ontogeny are maternal holdovers that the mother produced and deposited into the egg's cytoplasm. Much of the primary activation in the zygotic genome itself occurs after a ball of 4–16 preembryonic cells has formed. After further rounds of cellular division, a fluid-filled cavity forms within the growing cell ball. This hollow sphere, known as a *blastula*, consists initially of approximately 100 cells. To one side of its central cavity is a knob of undifferentiated cells, the inner cell mass from which embryonic stem cells are derived. Eventually, these cells give rise to all of the approximately 260 different cell types that make up each person's tissues, organs, and all other body parts (including gonads, which are destined to produce gametic cells of the potentially immortal germ line).

Another key event in pregnancy occurs when the *conceptus*, now called a *blastocyst*, implants in the mother's uterine wall during the second week after fertilization. Differentiation of the embryo's body parts (and the pregnancy itself) then begins in earnest. For example, by week four an embryonic heart takes shape and begins to beat. One week later, the embryo reaches a length of about one-third of an inch. By week eight, rudimentary precursors to all adult body structures are present, and the developing individual is termed a *fetus.* By the end of the first trimester (12 weeks), the 2-inch-long fetus becomes recognizably human (as opposed to another animal). By the end of the second trimester (at 24 weeks), the fetus is almost a foot long, and at 37–40 weeks the mother enters labor, and a baby is born.

The birth or delivery of a child represents the transition from the highly intimate bonds of internal gestation to what is thereafter a physical separateness of offspring and mother. Clearly, pregnancy is a major happening that directly affects a woman's life much more so than a man's. From an evolutionary perspective, pregnancy is yet another of the many pronounced reproductive asymmetries between the sexes that stem ultimately from anisogamy and can make procreative dramas so poignant in sexually reproducing species (table 5.1).

Although no book on pregnancy would be complete without at least some discussion of viviparity in *Homo sapiens* (fig. 5.1), I keep this chapter rather short for several reasons: (a) It would be presumptuous of me to expound unduly on a biological phenomenon about which nearly 50% of humanity (women) have far more personal knowledge than I do; (b) broader evolutionary ramifications of mammalian (including human) pregnancy are detailed in chapter 6; and (c) I prefer here just to introduce some fascinating peculiarities and oddities of human pregnancy, in both fact and fiction.

The mythologies and religions of most human cultures are replete with stories of varied biological phenomena that are often associated with real-life pregnancies. What follows are merely a few examples involving either the gods and heroes of ancient Greece (mostly from a review by Iavazzo 2008) or the folklore of Native Americans (mostly from Niethammer 1977).

TABLE 5.1   Reproductive disparities between human males and females.

Shown are merely a few examples of how dramatically men and women can differ in features related to reproduction and pregnancy. In the ratio column, each "1" applies to the male sex and the other number applies to the female sex.

Feature	Male	Female	Ratio
approximate size of a single gamete	$600\,\mu m^2$	$500{,}000\,\mu m^2$	1:1,000
lifetime number of mature gametes	10 trillion	400	250 billion:1
interval between erotic thoughts[1]	5 minutes	4 hours	50:1
minimum time invested in a pregnancy	10 minutes	9 months	1:40,000

[1]This estimate is merely a guess.

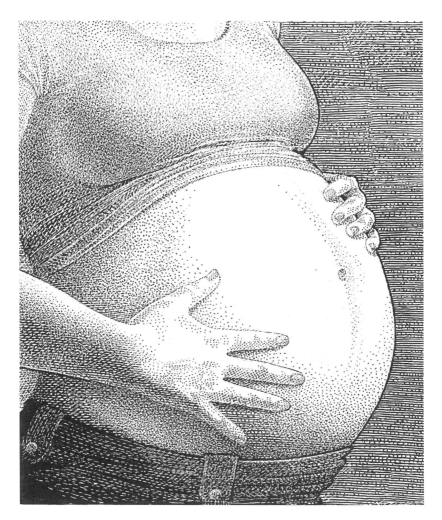

FIGURE 5.1   The beauty and burden of human pregnancy.

Conception and Contraception

Most human societies have recognized at least some general connection between sexual intercourse and pregnancy, but a clear understanding of fertilization and embryonic development is a surprisingly recent achievement in human history. In several Native American tribes, including the Apache, the blood (semen) from a man that enters a woman during coitus was thought to be opposed by the woman's blood, such that repeated lovemaking was required to build up the supply of male seed within a woman and thereby

---

Factoid: Did you know? As a fetus, each human female has about seven million potential eggs (primary oocytes), but only 350 of these survive to reach maturity in a grown woman, and only a few of the latter may become fertilized and eventuate in a successful pregnancy.

---

overcome her inherent resistance to pregnancy. The Hopi tribe of northern Arizona went one step further by supposing that continued sex throughout the pregnancy was important for fetal growth in much the same way that continued irrigation promoted the growth of their crops. On the other hand, women of the Fox tribe in Wisconsin generally abstained from sex during pregnancy for fear that their babies otherwise would be born "filthy."

With respect to overcoming problems of infertility, women in various Native American tribes tried all sorts of remedies: rituals, such as praying or placing little girls' clothing on a special "baby hill"; consultation with older women or with a wise medicine man; and numerous dietary gimmicks ranging from eating red ants to drinking boiled water saturated with rat urine and feces (an extreme practice reported in the Havasupai tribe living near the Grand Canyon).

Understandably, Native American women showed similar inventiveness in their ritualistic and other efforts to achieve exactly the reverse outcome: birth control (box 5.1). Indeed, quests for workable methods of birth control and its antithesis (fertility enhancement) have preoccupied nearly all human societies, cultures, and religions (Maguire 2003). Among native North Americans, some methods employed to avoid pregnancy include the following: brewing abortive potions from various plants in the local environments, as did many women in the Quinault and Salish tribes; licking the powder from dried and pulverized ground squirrels, as did some Havasupai women; and urinating on ant mounds or willing persistently that a child not be conceived, as did Yuma women who sought to avoid pregnancy.

BOX 5.1   Birth Control

Technically, the phrase "birth control" can refer to any method used either to prevent syngamy or to interrupt a pregnancy at any stage before parturition. It thus can include any of the following: any contraceptive device (such as a condom or diaphragm) or substance (such as a spermicidal gel) or behavior (such as coitus interruptus or withdrawal) or medical procedure

(continued)

BOX 5.1   (*continued*)

(such as a male vasectomy) that in effect foils the fertilization process; any contragestive approach (such as a physical intrauterine device or hormonal intervention) that prevents implantation of the blastocyst in the womb; or any abortion of a pregnancy that already is under way. To the extent that a birth control method is effective, it can give a couple some degree of pro-active control (other than sexual abstinence) over reproduction and family planning. Little wonder that women (and men) throughout the ages have sought to discover and employ effective birth-control methods. To the ancient Romans and Greeks, this often meant ingesting particular plants (such as pennyroyal or willow) or using intravaginal potions (such as mixtures of honey and acacia gum) that were thought to have contraceptive properties. Early texts in Islam and Christianity indicate that women have long used a variety of vaginal suppositories ranging from rock salt to elephant dung in their efforts to prevent unwanted pregnancies. Even some men occasionally did their part by using all sorts of contraceptive devices intended to short-circuit ejaculations or otherwise block sperm transfer to their mate.

In the modern era, several effective barrier methods are available for birth control, as are various ingestible, injectable, or implantable methods for delivering a range of hormones that may act either as contraceptives or contragestives. Most of the latter are designed for use by women, but attempts also continue to develop an effective oral or other contraceptive drug for men. Of course, modern medicine has also made abortion a much safer (although no less socially controversial) procedure of last resort.

Extracorporeal Fertilization

In 1978, the world witnessed the birth of the first "test-tube baby," Louise Joy Brown. Louise had been conceived by an assisted reproductive technology (ART) known as in vitro fertilization (IVF). In this then-revolutionary method, an ovum and a sperm are united in a petri dish (rather than in the woman's reproductive tract via coitus), and the resulting product is then implanted in a woman's uterus, where gestation of the embryo and fetus proceeds as usual. For many couples IVF proved to be a godsend because it can bypass numerous sources of infertility such as sperm immotility or even structural damage to a female's reproductive tract (especially if the IVF embryo is implanted and then incubated in the womb of a surrogate mother, as often is done). By 1995, nearly 150,000 IVF babies had been born, and the number of such babies today exceeds three million worldwide. Nowadays IVF is just one of several ARTs that can help couples (or even singles) produce babies in nonconventional ways (Avise 2004a).

Unconventional conceptions and ARTs were anticipated in Greek mythology. Indeed, two of the ancients' most famous female protagonists—the goddess Aphrodite and the beautiful Helen of Troy—both arose by nonstandard means. Aphrodite's story begins with the castration of Uranus (the father of the gods) by his son Cronus, while Uranus slept with the goddess Night. Cronus then threw the severed genitals into the Mediterranean Sea, where they formed a frothy brew from which Aphrodite (from *aphros*, meaning "sea foam") arose in what might be deemed an extracorporeal type of fertilization or pregnancy. Aphrodite later became the irresistible (aphrodisiacal) goddess of love and sexual rapture.

Helen of Troy's birth can be interpreted as an ancient example of surrogate motherhood (and perhaps of ovoviviparity, too). Helen was the daughter of Zeus and either Leda or Nemesis. The maternal ambiguity arises as follows. By one account, Nemesis initially refused Zeus's sexual advances by transforming herself into a goose. When Zeus discovered this trick, he transmogrified into a swan and slept with Nemesis. What resulted from this union was a blue egg that Leda later recovered and put into her vagina to keep warm. Thus, when Helen finally emerged from the inserted egg, her surrogate mother may have been the ovoviviparous Leda, but her biological or genetic mother apparently was Nemesis.

Superfetation

Superfetation is the co-occurrence of two or more embryos at different developmental stages in a pregnancy because fertilization took place on separate occasions within the mother's body, implying either that the embryos had different sires or that their mother's ova were fertilized at different times by the same male. Although superfetation is common in some animals, including particular fish species (see box 2.2), in humans it is an extremely rare occurrence at best, with fewer than a dozen (mostly anecdotal) reports of such happenings. In one such media-dramatized case, an Arkansas woman purportedly became pregnant twice, two weeks apart. In theory, superfetation in humans would normally require a conjunction of two rare events within the female body: continued ovulation after the first conception and successful implantation of a second fertilized egg at an atypical gestational site (see the later section on ectopic pregnancy). This type of conjunction is statistically highly unlikely because the joint occurrence of two or more rare events is equal to the product of their separate probabilities.

Factoid: Did you know? *Uterus didelphis* is a rare embryological mal-formation that results in the duplication of portions of a woman's repro-ductive tract, such that her uterus exists as a paired organ rather than the standard singleton. This condition apparently underlies some of the reports of superfetation in human twin pregnancies (Williams and Cummings 1953; Dorgan and Clarke 1956). For example, in one case a woman with uterus didelphis delivered twins with a birth interval of more than two months (Nohara et al. 2003).

Exactly how often bona fide superfetation occurs in *Homo sapiens* remains unclear, but what is undeniable is that superfetation took place in Greek my-thology. One such case involved the twin pregnancy of Hercules and Eficles, both born of Alcmene (the princess of the Greek kingdom of Mycenae and the wife of Amfitryon, the king of another Greek city-kingdom). This all came about because Alcmene had an illicit affair with Zeus on the same long night that she also slept with her husband, such that the two boys were conceived of different fathers. Presumably this explains why the twins had such differ-ent personalities. An analogous story applies to Leda's twin pregnancy of Castor (son of Zeus) and Pollux (son of the king of Sparta). Apparently Leda, too, displayed a high libido and was rather promiscuous.

Twinning or Multibirth Pregnancy

In modern human populations, approximately 1% of pregnancies result in the birth of twins, and in about one-third of those cases the twins are gene-tically identical (i.e., they are clonemates) (Bulmer 1970; MacGillivray et al. 1988). Identical twins, triplets, quadruplets, and so on are invariably the same sex (barring developmental anomalies). They arise via a phenomenon known as *polyembryony*, wherein a single fertilized egg mitotically divides into two or more genetically identical cells or sets of cells, each of which initiates embryo-genesis in the mother's womb. Thus, all polyembryonic offspring in a preg-nancy share a unique combination of paternal and maternal genes that were joined in a single union of sperm and egg. Monozygotic or identical twins are to be distinguished from dizygotic (nonidentical) twins, which arise from separate unions between sperm and oocytes. Genetically, dizygotic human twins (also called fraternal twins) are as different from one another as are any other full siblings (brothers and sisters from a woman's separate pregnancies with the same sire). Multiple offspring (such as triplets) within a single hu-man pregnancy sometimes include both fraternal and identical individuals.

Factoid: Did you know? A few well-documented human pregnancies have involved as many as eight babies born alive and more than a dozen embryos carried in utero for at least a short time. In recent years, multiple embryo transfer and the use of fertility drugs (which stimulate multiple eggs to mature in an ovulatory cycle) have greatly increased the frequency of human multiple births.

Extremely rare instances of the delivery of more than two monozygotic individuals are known in human pregnancies (Dallapiccola et al. 1985; Markovic and Trisovic 1979; Steinman 1998). Occasional or sporadic instances of identical twins have also been recorded in several other vertebrate species, including cattle (Ensminger 1980), pigs (Ashworth et al. 1998), deer (Robinette et al. 1977), whales (Zinchenko and Ivashin 1987), various birds (Berger 1953; Olsen 1962; Pattee et al. 1984), and fish (Laale 1984; Owusu-Frimpong and Hargreaves 2000). However, the phenomenon of sporadic polyembryony might be underreported for many sexual species because, to my knowledge, no one has searched methodically—using suitable genetic markers—for polyembryonic (clonemate) progeny in nature.

On rare occasions, the bodies of identical twins become joined in utero and might later be delivered as "conjoined twins" (also known as "Siamese twins"), whose bodies are partly fused. Conjoined twins, who usually arise by the partial fission of a fertilized egg, share the same embryonic membranes and placenta during gestation. Today, conjoined human twins can sometimes be separated by surgical procedures (depending on the placement and extent of their melded body parts). However, some conjoined twins grow and spend their lives together in a partially welded condition.

Twins of various sorts were also a recurring theme in Greek and other mythologies. One such example mentioned earlier involved Hercules and Eficles, both born of Alcmene but sired by different fathers. In this fictional case, Hercules and Eficles would have been "half-sib twins" rather than "full-sib twins" because they shared only one parent. Another fictitious example of a half-sib twin pregnancy (i.e., superfetation) involved Leda's gestation of Castor and Pollux (mentioned earlier). Apollo (the sun god) and Artemis (the moon goddess) are a more conventional pair of twins from Greek mythology.

Factoid: Did you know? In humans, the incidence of conjoined twins is about one such pair per 50,000–100,000 live births.

Many other mythological twins likewise have dualistic natures or split personalities. Some examples include the Egyptian figures Geb (the earth god) and Nut (the sky goddess); the twins Ahriman (the spirit of evil) and Ahura Mazda (the spirit of good) of Zoroastrian mythology; and the twin brothers Kuat (who became the sun) and Iae (the moon god) in the creation mythology of the Xingu people of Brazil.

For many Native North Americans, twins had a special significance both in mythology and in peoples' mortal lives. In an example of mythological twins, some northeastern tribes believed that the creator god Gluskap had an evil twin, Malsum, the ruler of demons. With regard to more concrete examples from daily life, some tribal societies supposed that a woman's birth of twins evidenced a special potency or virility of her mate or that twins perhaps resulted from too much sex. And in some Native American cultures, women who wished not to conceive twins often avoided eating paired fruits, such as double almonds or split bananas.

Ectopic Pregnancy

This unusual phenomenon—implantation of the embryo outside the normal uterine cavity—is a contributor to spontaneous abortion and an important cause of maternal mortality during the first trimester of human pregnancy (Khan et al. 2006). A fallopian tube, which connects each ovary to the uterus, is the most common site for an ectopic pregnancy, but in rare circumstances other known locations have been a woman's ovary, her general stomach area (Tait 1880; Stromme et al. 1959), and her cervix (the constricted lower end of her uterus). The developing cells of the embryo cannot survive for long in these nonstandard locations, so real-world ectopic pregnancies almost invariably fail.

Ectopic pregnancies are not necessarily lacking in the world of Greek mythology, however. Athena (the goddess of wisdom, war, arts, industry, justice, and skill) was live-born from a truly bizarre form of ectopic pregnancy. Athena was the daughter of Zeus and his first wife, Metis. During Metis's pregnancy, Zeus heard a prophecy that his child would someday slay him in order to inherit the throne. To avoid this fate, Zeus ate Metis, only to find himself suffering from a severe headache exactly nine months later. To alleviate this

---

Factoid: Did you know? Fallopian tubes get their name from Gabriele Falloppio, a prominent Italian anatomist and physician of the 16th century. He worked mostly on the anatomy of the human head and ear, but he also studied the reproductive organs of both sexes.

pain, Prometheus (the Titan god of forethought) opened Zeus's head with an axe, and out popped a healthy Athena. Zeus's head had provided a serviceable substitute womb for this strangely ectopic pregnancy!

From Preterm Labor to Prolonged Pregnancy

By definition, any onset of human labor prior to 37 weeks of gestation can lead to the "preterm birth" of a premature baby (sometimes insensitively called a "preemie"). In the United States, preterm birth occurs in about 12% of pregnancies and is a leading cause of perinatal ("near-birth") mortality (Goldenberg et al. 2008). Worldwide, "prematurity" yields about 500,000 neonatal deaths per year (lung failure is a common proximate causal agent because the lungs are among the slowest organs to develop in a fetus). All else being equal, the survival rates of premature babies tend to increase in direct proportion to the duration of prelabor gestation. However, premature infants who receive heroic medical care (including attachment to a respirometer) stand a reasonably good (50%) chance of survival even if they have gestated internally for as little as 24 weeks (Kaempf et al. 2006).

Prolonged pregnancy (Higgins 1954) lies at the opposite end of the spectrum from preterm labor. Usually a prolonged pregnancy in humans is defined as any gestation that exceeds 41–42 weeks. Without intervention (i.e., via the induction of labor or a cesarean section), 7–18% of all pregnancies in the United States reach these impressive temporal milestones.

In Greek mythology, a famous story of a preterm birth in one pregnancy and a delayed birth in another involved the pregnancies of Alcmene, when

---

Factoid: Did you know? The shortest-term premature human baby on record had a gestation of only 21 weeks and 6 days.

Factoid: Did you know? Compared to other primates, gestation times in humans have evolved to be extraordinarily short relative to neonatal body size, probably because of our need to deliver large-headed babies through a narrow birth canal. Furthermore, scientists recently discovered one of the "birth-timing" genes responsible for the relatively short gestations in *Homo sapiens* (Haataja et al. 2011).

she bore Heracles (one of the greatest of the Greek heroes), and of Nikippi, when she bore Eurystheas (who would later rule the city of Argos). The unusual durations of these two pregnancies came about as follows. During Alcmene's pregnancy, Zeus (the likely father) announced to the gods that the first child born to the family of Perseas (the grandfather of Alcmene) would become the king of Argos. Hera (Zeus's wife) understandably became disturbed when she learned of this promise, so she asked Eilytheia (the goddess of labor) to help her thwart Zeus's plans. Eilytheia complied by delaying Alcmene's labor while at the same time greatly hastening that of Eurystheas. As a result of these two temporal adjustments, when Zeus's oath was fulfilled, it was indeed Eurystheas (not Heracles) who eventually became the ruler of Argos.

Leto (another of Zeus's many lovers) is another example of prolonged pregnancy in the tales of the ancient Greeks. When a jealous Hera learned of Leto's pregnancy, she sent a big snake (a python) to chase Leto and thereby thwart Leto's attempts to find a safe birthing site. This long and eventful chase across much of Greece greatly delayed Leto's birth of Apollo and Artemis and thus can be considered the root cause of Leto's prolonged pregnancy.

Spontaneous Abortion

A spontaneous abortion or miscarriage is usually defined as the expulsion (and death) of a human embryo or fetus before approximately the 22nd week of gestation. Particular instances may have a genetic etiology (frequently involving difficulties in chromosomal replication in the embryo; Plachot 1989), or they may be due to physical trauma or to biochemical or physiological difficulties during a pregnancy. Approximately 10–50% of human pregnancies eventuate in a clinically apparent miscarriage, but these statistics are misleadingly low because most miscarriages occur very early in a pregnancy and often are unbeknownst even to the mother. For example, one scientific study found that 70% of conceptuses were lost within the first 12 weeks postfertilization (Edmonds et al. 1982), and another study (Roberts and Lowe 1975) estimated that 78% of human conceptions never come to a full-term fruition. Thus, human pregnancy far more often leads to the death rather than to the successful birth of a new person.

Factoid: Did you know? Human mothers have remarkably effective physiological screening mechanisms that endogenously examine and often sift out (spontaneously abort) genetically defective embryos (Forbes 1997).

In general, a successful C-section can be defined as the survival of both mother and child for at least 1 month after the CS was performed. By this criterion, successful CSs were extremely rare before the introduction of anesthesia, antisepsis, and other premodern surgical procedures of the 19th century. Nevertheless, occasional reports persist from earlier times of partly successful CSs (in the sense that the baby may have survived). For example, Bindusara (an early emperor of India) was reportedly delivered by CS just after his pregnant mother had died by consuming poison. Today, cesarean sections are standard practice in many countries. For example, nearly one-third of all babies are now delivered by CS in some parts of the United States and in particular regions of other countries such as Italy and Brazil.

Factoid: Did you know? Perhaps the world's oldest pregnant woman—Omkari Panwar—gave birth by cesarean section to twins at the age of 70. (However, lack of a birth certificate for Panwar makes the precise age of her pregnancy less than certain.)

Long before modern medicine, cesarean sections were the stuff of legend. In Greek mythology, Prometheus's use of an axe to cleave open Zeus's head (mentioned earlier) could be deemed a rather literal example of a "cesarean section" of a highly ectopic pregnancy. A more conventional type of CS from that era involved the birth of Asklepios, the Greek god of medicine and the son of Apollo. Apollo himself performed this CS after he learned that his beloved nymph, Coronis, had been unfaithful. Enraged, Apollo arranged for the murder of Coronis, but then, in an act of compassion, he removed his son from the dead body of Coronis even as she lay on the funeral pyre.

"Abdominal delivery" via CS was also a recurring topic in several of the world's prominent religions (van Dongen 2009), including Buddhism (Buddha was born from the right flank of his mother, Maya), Judaism (delivery by a cut in the abdomen was mentioned in the Jewish law, Mishnah), and Christianity (where by some accounts the Antichrist was born via CS and thereby symbolizes the destruction of both mother and child). (Given all the biological mishaps that can occur before, during, and shortly after embryonic gestation in humans, pregnancy indeed might seem like a devil's playground.) Finally, even Shakespeare used the concept of cesarean section in one of his plots by invoking this poetic phrase: "for none of woman born shall harm Macbeth." Thus, Macbeth believed he was invincible until, in the end, he is killed by a man (Macduff) who "was from his mother's womb untimely ripp'd" (presumably by cesarean section, thus rendering Macbeth mortal after all).

Precocious Puberty

This phrase refers to an early onset of puberty for any reason: normal extremes in the wide standard variation of human development or, perhaps in other cases, to overt hormonal imbalances triggered by a pituitary tumor or by some other trauma to a person's brain or gonads. Typically, puberty is considered at least somewhat early if it occurs in a child who is younger than about 8 or 9 years of age. Otherwise, the time of first menstruation (menarche) for girls in developed Western countries is about 13 years. The onset of menstruation (box 5.2) is an especially important component of female puberty because it signals the beginning of a woman's effective fecundity. Precocious puberty can make a girl capable of conceiving and giving birth while still very young.

BOX 5.2   Menstruation (Menses)

This is the periodic flow of blood and mucosal tissue from the uterus to the outside of the body via the vagina. In humans, the material expressed normally consists of a few teaspoons of blood plus remnants of the endometrium (the lining of the uterus). Menstruation in a woman occurs on a more or less monthly schedule, lasts 2–8 days, and is often accompanied by cramps (dysmenorrhea), caused mostly by contractions of uterine muscles. Each menstrual period is part of a monthly cycle of hormonal and reproductive changes, during which a haploid egg cell completes its maturation in the ovary and is shed (ovulated) into the oviduct, where it temporarily becomes available for fertilization. Ovulation usually occurs midway through the monthly cycle (i.e., about 2 weeks after and 2 weeks before successive menstruations).

Overt menstruation, which is rare in the biological world, is confined mostly to humans and some of our closest evolutionary relatives, including chimpanzees. Instead, most other placental mammals have *covert menstruation*, in which a female reabsorbs the endometrium at the end of her reproductive cycle. Anthropologists have speculated on this evolutionary difference, and one has suggested that females with covert menstruation save considerable energy by not having to rebuild the uterine lining for each fertility cycle (Strassmann 1996). But why then do humans have overt menstruation? One hypothesis is that prolonged fetal development in *Homo sapiens* requires a thick and highly developed endometrium that is too difficult for a female to reabsorb completely (Strassmann 1996). Furthermore,

(continued)

BOX 5.2   (*continued*)

appreciable energetic costs might be associated with having to maintain a uterine lining continuously.

Menarche, in effect marks the beginning of a woman's fertility, which later ebbs and eventually ends with the cessation of menstruation during menopause. Also, menstruation may be suspended temporarily (a condition known as *amenorrhea*) during otherwise fertile portions of a woman's lifespan, such as pregnancy, the period immediately following childbirth, and episodes of exceptional physical or emotional stress or ill health.

Many cultures and religions have special traditions regarding menstruation. For example, orthodox Judaism and Islam discourage intercourse during a woman's "period," and some traditional societies construct "menstrual huts," where women must sequester themselves at the appropriate times.

Factoid: Did you know? The youngest well-documented mother (and successful pregnancy) on record involved a Peruvian girl, Lina Medina, who gave birth by cesarean section at the age of 5 years and 8 months (Escomel 1939). She was estimated to have begun menstruating at 1–2 years of age and had prominent breast development by age four.

"Delayed puberty" resides at the opposite end of this temporal spectrum. Although precise definitions of the phenomenon are difficult because the age of puberty shows considerable variation even in healthy populations, puberty might be considered exceptionally late or even pathologically delayed if a girl fails to experience menarche by the age of 18 or if a boy shows little testicular development by about age 20. Among the possible causes are exogenous factors such as malnutrition or disease and numerous endogenous factors, including genetic disorders or the body's inappropriate production of or response to sex hormones.

Standard Vaginal Delivery

When discussing human pregnancy in an evolutionary framework, perhaps the most widely appreciated fact is the difficulty of childbirth due to the large size of a fetus's head relative to the birth canal. By favoring an increased mental capacity in our evolutionary ancestors, natural selection seems to have pushed the size of the human brain and cranium nearly to the physical limit, as any woman who experiences the pain of labor's second stage (fetal expulsion) is likely to attest. Indeed, despite the remarkable dilation of the woman's reproductive tract (notably the cervix, which is the gateway from the uterus to the vagina) that occurs during the first active phase of childbirth, an obstetrician may later perform an *episiotomy* or a *perineotomy*—a surgical incision in the woman's posterior vaginal wall that enlarges the opening and allows the baby to squeeze through. In addition to the physical challenges of childbirth, the delivery phenomenon has psychological consequences for the mother that can range from great elation to postpartum depression. The relative immaturity and helplessness of human babies compared to the offspring of many other primates are further testimony to the evolutionary premium of keeping babies small enough to pass through the birth canal despite the competing selection pressures for larger heads and brains. As is true in so many facets of human life, successful reproduction under an evolutionary view is an amalgam of oft-competing interests and tradeoffs between opposed selection pressures.

With respect to the difficulty of childbirth, humans are exceptional but hardly unique. In many other mammalian species such as giraffes and other *ungulates* (hoofed animals), females seem to experience considerable discomfort as they give birth to large or otherwise cumbersome infants. On the other hand, childbirth in many other mammals seems downright effortless. In some marsupials, for example, mothers deliver tiny babies, who then continue much of their development in an external *marsupium*, or pouch. But bears (Ursidae) provide even better examples of carefree childbirths. Consider, for example, the black bear (*Ursus americanus*), in which a 300-pound pregnant female gives birth during hibernation (effectively while asleep) to infants so tiny that each could fit inside a teacup. During the winter months, as mom continues to snooze in her den, her cubs suckle and grow from the nutrition that she provides by transforming her thick body fat to milk.

In humans, another component (the third, or final, stage) of a standard vaginal delivery typically occurs within 15–30 minutes of childbirth. It is the expulsion of the afterbirth (i.e., the remnants of the placenta). In Egyptian mythology, the redness of the dawn sky after a dark night was sometimes

---

Factoid: Did you know? Maternal mortality is defined as the death of a woman during pregnancy or in the 42 days postpartum due to causes directly or indirectly associated with the pregnancy. In recent decades, maternal mortality worldwide has averaged about 40 deaths per 100,000 human live births, or more than half a million maternal deaths per year (Hill et al. 2007).

> **Factoid: Did you know?** In many other mammalian species, the mother, after giving birth, bites through the umbilical cord and then eats it and the placental remnants—a feeding behavior known as *placentophagy*.

interpreted as a bloody afterbirth that accompanied the birth of the sun god with each new day.

The Placenta

During a pregnancy, the human placenta (from the Latin "plakóeis," meaning flat cake) is a disk-shaped organ on the mother's uterine wall that attaches to a ropelike structure (the umbilical cord) that serves as a conduit not only for delivering nutrients from the mother to the child but also for exchanging gases (oxygen and carbon dioxide) and eliminating embryonic wastes.

The mammalian placenta is a crucial physical link between embryo and mother. It arises early in a pregnancy as follows (fig. 5.2). After the new zygote has undergone several mitotic cell divisions, the resulting cells separate into two genetically identical subpopulations: the inner cell mass, which eventuates in the embryo, and the trophectoderm, which will form the placenta. Thus, from a genetic perspective, cells of the placenta are identical to those of the prospective fetus, so in this respect the placenta can be deemed the embryo's identical twin since both share the same cramped womb. From a physiological and biochemical perspective, however, the placenta has both fetal and maternal components that were generated after the blastocyst implanted in the mother's uterine wall. The placenta grows throughout a pregnancy and produces a succession of reproductive hormones (some constructed from imported maternal precursor molecules) that help to orchestrate the gestational process and direct the embryo's development (Bonnin et al. 2011). Indeed, some of these hormones (such as placental lactogen or chorionic somatomammotropin in some species) have effects that extend beyond parturition, such as by preparing the mother for postpartum lactation.

Although the placenta is at the maternal-fetal interface, the maternal and fetal bloodstreams do not intermingle directly. Instead, microcapillaries from

> **Factoid: Did you know?** The human placenta is nearly 1 foot long, 1 inch thick at its center, weighs about 1 pound, and is maroon in color. The umbilical cord, to which it connects, is about 2 feet long and terminates at the fetus's "belly button," the navel.

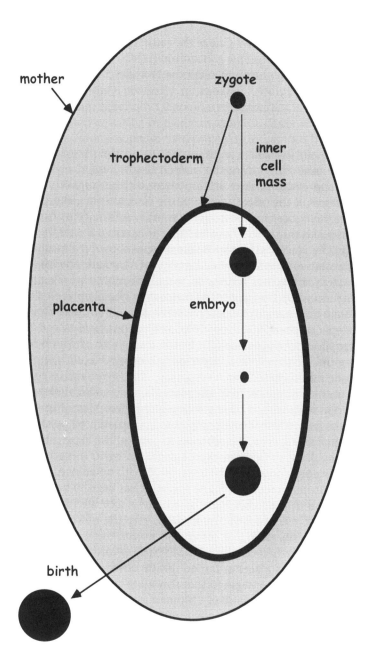

FIGURE 5.2 The ontogenetic origin of each mammalian placenta and its associated embryo (after McKay 2011). From a genetic perspective, the two can be thought of as identical twins (see the text).

Factoid: Did you know? A few "bona-fide male pregnancies" widely reported in the media have involved transgender humans who were born with functional ovaries and other female body parts but after being impregnated consider themselves pregnant transgender men.

society. In the 1994 comedic film *Junior*, a very manly scientific researcher, Alex Hesse (played by Arnold Schwarzenegger, recent governor of California), becomes pregnant after accidentally taking an experimental drug designed to reduce the chances of a woman having a miscarriage. Alex soon becomes emotionally attached to his unborn baby, which gradually grows to considerable size within his "Arnie's womb."

Can a human male truly get pregnant in real life? As the promotional ads for *Junior* proclaim, "Nothing is inconceivable." Indeed, in 1992 the tabloid world was shocked by media reports (complete with convincing pictures) of a very pregnant Taiwanese man (a Mr. Lee) who had purportedly volunteered to have an IVF embryo implanted into his abdominal cavity, with the expectation that a full-term baby would later be delivered by cesarean section. Unfortunately (or perhaps fortunately), the report and the doctored photos turned out to be nothing more than an elaborate hoax. Still, the account prompted much speculation about what might theoretically be possible in the age of assisted reproductive technologies. Assuming that many biological and technological hurdles might someday be overcome, it is perhaps conceivable that some sort of ectopic implantation could produce a "pregnant" man, who carries an embryo internally, at least for some time. However, relatively few men may be eager to explore such possibilities.

Finally, many science-fiction writers have envisioned strange reproductive worlds. For example, in Ursula K. Le Guin's *The Left Hand of Darkness*, anyone can become pregnant, whereas in Marge Piercy's *Woman on the Edge of Time*, no one becomes pregnant, and, instead, all human gestations transpire in artificial wombs.

SUMMARY

1. Pregnancy is a prominent feature of human existence not only in obstetrical practice but also in many of our mythologies and cultural traditions. Much is mechanistically understood about the standard course of real-life events during a human pregnancy, from its beginning with conception and implantation to its conclusion at parturition of the child and its afterbirth twin.

2. Much is also known about various pregnancy phenomena that are somewhat less standard. Some of the atypical expressions of pregnancy that have been documented in humans include the following: extracorporeal fertilization (syngamy at a site other than the normal location within the female's reproductive tract); superfetation (the co-occurrence of two or more embryos at different stages of development); multibirth pregnancy (the bearing of two or more same-stage offspring that may be either genetically identical [monozygotic] or different [multizygotic or fraternal]); ectopic pregnancy (implantation of an embryo outside the normal uterine cavity); preterm labor, which may yield a "preemie" baby; prolonged pregnancy, which lasts beyond about 41–42 weeks; spontaneous abortion, which terminates a pregnancy prematurely; cesarean section, the delivery of a baby by means of an incision in a mother's abdomen and uterus; precocious puberty, wherein a female may become pregnant at an unusually young age; and male pregnancy, wherein an apparent man (or more likely a transgender person) becomes impregnated.

3. All of these and other pregnancy-related phenomena are also richly represented in the mythologies, cultural beliefs, and religious practices of human societies, giving further testament to the central position of pregnancy in peoples' thoughts throughout human history. Many fictitious tales about gestation, though sometimes wildly exaggerated, find recognizable analogues in some real-life human pregnancies.

4. Among many other subjects that arise routinely in discussions of human pregnancy are the following: the placenta and umbilical cord, which physically connect mother and fetus; menopause, which normally ends a woman's potential to become pregnant; various methods of birth control; and menstruation (menses) and menarche (the time of a girl's first menstrual period). In many human societies past and present, all of these subjects are deeply engrained in cultural superstitions, folklore, customs, and rituals.

Evolutionary Ramifications of Pregnancy

Several forms of selection motivate and in turn become amplified by the gestational phenomena described in part I. The chapters in part II address these evolutionary forces. Chapter 6 explains how natural selection can affect mammalian (including human) pregnancies; chapter 7 discusses how sexual selection emanates from and affects piscine pregnancies; and chapter 8 reexamines these topics from the vantage point of comparative evolution. The chapters in part II have the following overarching themes: (a) various forms of pregnancy represent extreme expressions of gender-biased parental care that in effect amplify the inherent intersexual (male versus female) tensions that originated millions of millennia ago during the evolutionary emergence of anisogamy and sexual reproduction; (b) pregnancy (like many other aspects of ontogeny) entails compromises and trade-offs between opposing selective forces; and (c) pregnancy from an evolutionary perspective is as much about conflict as it is about unbridled cooperation among mother, child, father, and copies of their respective genes.

CHAPTER SIX

Natural Selection During Mammalian Pregnancy

Pregnancy unleashes powerful forms of natural selection during the gestational phase of the mammalian life cycle. From the outset, however, I want to emphasize that not all expressions of mammalian pregnancy necessarily register adaptations shaped by natural selection. Thus there is no need to invoke adaptive justifications for all empirical facets of mammalian pregnancies. Indeed, chapter 5 noted that human pregnancy is often maladaptive (even lethal) to the participants. Here we take a broader evolutionary look at how natural selection both shapes and can be directed by mammalian-style viviparity. I first discuss natural selection's likely influence (or sometimes lack thereof) on the evolution of three common gestational phenomena in mammals: (a) delayed implantation, which indeed demands an adaptive explanation; (b) sporadic polyembryony, which calls for no adaptive interpretation; and (c) dizygotic twinning, some biological elements of which require adaptive explication and others probably do not.

Delayed Implantation

In some viviparous animals such as badgers (fig. 6.1), skunks, and wolverines, long intervals occur between fertilization and implantation of an embryo in the dam's uterus. In other words, in species with delayed implantation (DI), the developmental "tabs" for syngamy and implantation have evolved to be further apart along the temporal axis of ontogeny (see fig. 1.4). Delayed implantation is merely one subcategory of *embryonic diapause*, which is defined broadly as any temporary arrest in embryonic development regardless of the

FIGURE 6.1   American badger, *Taxidea taxus* (Mustelidae), a species that displays delayed implantation or embryonic diapause.

Factoid: Did you know? Although humans do not have delayed implantation, one artificial case of DI was described in *Homo sapiens* (Grinsted and Avery 1996). Following in vitro fertilization (IVF) and embryo transfer, a woman eventually gave birth, but her pregnancy in effect had begun after a 5-week delay following the original aspiration and IVF of her oocyte.

mechanism. Delayed implantation can extend the duration of a pregnancy, sometimes dramatically.

In nature, embryonic diapause is routine in almost 100 mammalian species representing 7 taxonomic orders (Renfree and Shaw 2000). Members of the family Mustelidae are of special interest because DI is highly developed in some mustelid lineages but absent in others (fig. 6.2). Delayed implantation is probably the ancestral condition in Mustelidae, in whom several

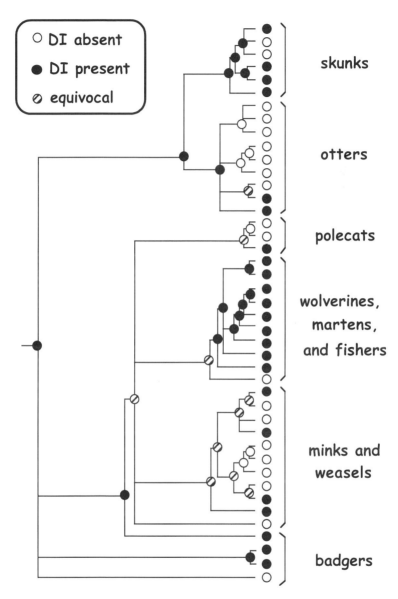

FIGURE 6.2   Phylogenetic distribution of delayed implantation or embryonic diapause along an evolutionary tree for mammalian species in the family Mustelidae (from Avise 2006).

> **Factoid: Did you know?** Mustelidae (from the Latin "mustela," meaning "weasel") is the largest and most diverse family of carnivorous mammals.

independent evolutionary switches between DI and non-DI are required to explain the phylogenetic arrangement of embryonic diapause among extant species (Lindenfors et al. 2003; Thom et al. 2004).

Given that DI evolved convergently on several independent occasions in mustelids and various other mammals, what (if any) is its adaptive significance? One compelling idea is that DI could enhance an animal's genetic fitness whenever natural selection favors a decoupling of the times of mating and parturition. This should be the case, for example, in environments where harsh winters and pronounced seasonality give a fitness advantage to any female who can mate in one season (e.g., the autumn) but delay giving birth until a distant season (typically the spring or summer) that may better facilitate the survival of her progeny. In many bears (Ursidae), DI occurs on a different schedule, which eventuates in a midwinter birth as the mother delivers her offspring while she hibernates. Thus, a generalized version of the adaptationist hypothesis is that DI could be beneficial in any biological or ecological setting in which the optimal periods for mating and birthing differ. During evolution, natural selection then acts to adjust the duration of embryonic diapause in each species so as to achieve a suitable match between reproductive biology and environmental demands.

Polyembryony

Polyembryony, or monozygotic twinning, is the production of two or more genetically identical offspring (clonemates) from a single zygote. Depending on the taxon, polyembryony may be either sporadic (occasional) or constitutive (when it occurs consistently in a species). Polyembryonic pregnancies in sexual organisms present an evolutionary enigma (Craig et al. 1995, 1997). Why would natural selection *ever* favor the production of polyembryos (as opposed to genetically diverse offspring) within a clutch? The famous evolutionary biologist George Williams (1975) analogized the mystery to the purchase of multiple lottery tickets with the same number even though no reason exists to prefer one number to another. In any polyembryonic pregnancy, the parents' entire evolutionary wager for the litter is placed on one cloned genotype. Furthermore, because that genotype was generated sexually and differs from those of both parents, it is functionally untested and ecologically unproven. The evolutionary paradox is that polyembryony appears to lack the

fitness advantages normally associated with either sexual or asexual reproduction. Standard sexual reproduction yields genetically diverse progeny, at
least some of which might fare well in new ecological circumstances, and
standard asexual reproduction (as in parthenogenesis) yields genetically
uniform progeny whose multilocus genotype survived field testing in the
parents. However, neither of these benefits would seem to occur with polyembryony. Thus, polyembryony at face value seems to combine some of sexuality's worst elements (the deployment of newly recombined and untested
genotypes) with clonality's worst elements (lack of genetic heterogeneity
within a brood).

Sporadic

As described in chapter 5, polyembryony in humans and many other vertebrates is rare and sporadic. Furthermore, because fraternal twins are usually
more common than identical twins, there is no compelling evidence that sporadic polyembryony per se offers fitness advantages during a multibirth pregnancy. Thus, although polyembryony in humans is scientifically interesting
and certainly has social and medical implications for affected families, it requires no special adaptive explanation, nor is it likely to have dramatic consequences for how natural selection operates in humans or other mammalian
species in which the phenomenon is rare and sporadic. The same sentiment
may not be true, however, for sexual species with constitutive polyembryony.

Constitutive

Armadillos in the genus *Dasypus* are the only vertebrates known to produce
polyembryonic litters consistently and apparently exclusively. Six extant *Dasypus* species reside mostly in South America, but the nine-banded armadillo
(fig. 6.3) reaches the southeastern United States. Each litter of *D. novemcinctus* typically consists of genetically identical quadruplets, although unusual
instances of twins, triplets, quintuplets, and sextuplets have been reported
(Newman 1913; Buchanan 1957; Galbreath 1985). In other species of the genus, standard litter sizes vary from two (*D. kappleri*) to four or eight (as in *D.
sabanicola*, *D. septemcinctus*, and *D. hybridus*) to as many as twelve in rare instances. Several other armadillo genera also reside in the Dasypodidae. The
closest relatives of *Dasypus* appear to be in *Tolypeutes* and *Cabassous*, two genera with species that usually produce only one offspring per pregnancy (Wetzel 1985). This taxonomic distribution of litter sizes indicates that constitutive
polyembryony in the *Dasypus* lineage probably arose from an ancestral condition of one armadillo pup per litter.

Polyembryony in *Dasypus* armadillos has long been suspected from indirect evidence: littermates are invariably the same sex, and multiple embryos

FIGURE 6.3   Nine-banded armadillo, *Dasypus novemcinctus* (Dasypodidae), a species that consistently gives birth to monozygotic quadruplets.

Factoid: Did you know? "Armadillo" is Spanish for "little armored one," in reference to the animal's leathery, plated shell, which serves as a protective covering.

in each pregnancy are encased in a single chorionic membrane. Not until late in the 20th century, however, was the clonal identity of armadillo littermates confirmed directly from molecular markers (Prodöhl et al. 1996, 1998) and other genetic evidence (Billingham and Neaves 2005).

How might constitutive polyembryony have arisen in armadillos, and should natural selection be invoked? One chain of argument about the evolution of obligate polyembryony proceeds as follows. In any sexual species, parents who produce nonclonemate offspring are hedging their bets by securing different genetic tickets in the reproductive sweepstakes. However, from each offspring's selfish perspective, a heavy parental investment in one specific genotype (its own and that of its polyembryonically identical siblings) is the preferred option. Thus, in sexual species with extended parental care of young, parents and their progeny typically have inherent conflicts of interest over optimal parental investment tactics (Parker 1985; Stamps et al. 1978; Trivers 1974). By this reasoning, polyembryony might be interpreted as a special case in which the offspring's preferred tactic in effect has won this evolutionary tug-of-war (Williams, 1975). The theoretical challenge then would be to elucidate particular ecological or ontogenetic conditions in which any genetic disposition for monozygotic twinning might invade the gene pool of a population otherwise engaged in sexual reproduction without poly-embryony (Gleeson et al. 1994). On the other hand, the tug-of-war metaphor might be entirely inappropriate.

Musings about the optimal genetic composition of a brood have given rise to several other hypotheses about how constitutive polyembryony might evolve in particular species. With respect to life-history features, conventional speculation has been that polyembryony might be favored (all else being equal) in species in which early stages of embryonic development are lengthy, pregnant females face unpredictable resource availabilities, and/or when sperm cells are in limited supply, because a case can be made in each such biological circumstance that polyembryony might be of benefit to both a pregnant female and her offspring. However, it seems unlikely that these conditions apply with any special force to armadillos as compared to the other vertebrate groups that have not evolved constitutive polyembryony. Thus, some other evolutionary explanation for constitutive polyembryony may be required.

Another general line of speculation is that polyembryony tends to evolve when parents have less information about optimal clutch size than do their offspring (Craig et al. 1997). Whenever progeny are in the best position to judge the quality or quantity of environmental resources that will be available to them, polyembryony could be favored because it might allow such littermates to adjust the extent of their clonal proliferation accordingly. This hypothesis was motivated by the observed tendency in invertebrate animals for polyembryony to be associated with endoparasitism (see box 6.1). However, it seems doubtful that *Dasypus* biology uniquely qualifies armadillos to have evolved polyembryony in this manner.

BOX 6.1   Polyembryony in Endoparasitic Invertebrates

Polyembryony is a regular occurrence in invertebrate animals in several phyla (Craig et al. 1997; Avise 2008). Many of these species show internal brooding or various other pregnancy-like phenomena. For example, bryozoans in the order Cyclostomata are small colonial marine animals that produce up to hundreds of genetically identical progeny (known as *gonozooids*) in specialized brood chambers (Reed 1991; Ryland 1970). A very different form of polyembryonic "pregnancy" is displayed by a freshwater

(continued)

BOX 6.1 *(continued)*

hydrozoan (*Polypodium hydriforme*; Cnidaria) that uses eggs of the fish species that it parasitizes as brooding chambers for its young.

Many other polyembryonic invertebrates are also *endoparasites* (Godfray 1994) (i.e., parasites that live within the body of a host organism). For example, a tiny wasp, *Copidosoma floridanum* (see the figure in this box), which parasitizes moths, provides another fascinating expression of both polyembryony and interspecific larval brooding (Strand 1989a, 1989b).

In this species, each female wasp oviposits one or two eggs in an egg of the host moth. After the host egg hatches and begins to develop into a caterpillar (a "heterospecific womb"), the wasp egg divides mitotically and produces hundreds or even thousands of polyembryos within the host (Grbic et al. 1998). A few of these clonemate wasps become soldiers that patrol inside the caterpillar's body to prevent subsequent invasion by other parasitoids (Cruz 1981; Giron et al. 2004; Hughes 1989). Other members of the brood eventually kill the caterpillar by eating their way out of its body. The wasp larvae then pupate on the corpse's skin. *Copidosoma floridanum* is just one of many polyembryonic parasitoid wasps, an invertebrate assem-

(continued)

BOX 6.1   (*continued*)

blage in which polyembryony has evolved convergently on multiple occasions (Craig et al. 1997).

Craig et al. (1997) interpreted the association between polyembryony and parasitism in many invertebrates as an evolutionary response to selection pressures related to a mother's inability to predict her optimal brood size. In the case of *Copidosoma floridanum*, for example, each female who oviposits into a host's egg would ideally "want to know" just how big the moth caterpillar (her progeny's food source) will become so that she can properly adjust the number of fertilized eggs that she deposits. But such information is presumably unavailable to her. Furthermore, the moth egg offers little space for a female wasp to oviposit a multiegg clutch (Hardy 1995). Given these special circumstances, both the mother wasp and her progeny might benefit if proliferation within the brood is delayed until the caterpillar stage of the host, and polyembryony is the only available postoviposition mechanism for achieving this outcome. A further potential advantage of polyembryony in this biological setting is that competition among clonemates should be minimized and collaborative behaviors rewarded via the intense kin selection that can come into play within the tight ecological quarters of the host caterpillar.

Interestingly, an analogous argument can be made for the adaptive significance of polyembryony in armadillos if the "uterine constraint hypothesis" (see the text) is correct (Loughry et al. 1998). For the parasitoid wasps, a tiny moth egg is the resource bottleneck that later expands into a spacious caterpillar, whose food-rich body can support the development of many wasp polyembryos. For the armadillos, a tiny uterine implantation site is the resource bottleneck that later expands into a spacious womb that can house and nourish multiple clonal littermates. Thus, for wasps and armadillos alike, polyembryony circumvents a temporary restriction on brood size by capitalizing upon what soon becomes a richly expanded ecological setting (caterpillar or uterus, respectively) for further embryonic development.

A third general hypothesis for the evolution of polyembryony entertains the notion of nepotism (i.e., favoritism toward kin) (box 6.2). As applied to armadillos, a kin-selection hypothesis for polyembryony would require that littermates help one another to build dens, find food, detect predators, groom parasites, defeat competitors, or otherwise benefit from mutual aid. If so, any genes underlying polyembryony might have been favored by kin selection if pronounced cooperation among clonal littermates led to higher mean rates of survival and successful reproduction of clonal littermates than was true for nonclone littermates. Alas, this appealing evolutionary hypothesis also appears inadequate to account for polyembryony in armadillos. Genetic analyses in conjunction with field investigations have revealed that weaned armadillos in nature seldom remain together long after birth, and even when pups are found together, they appear not to display nepotistic behaviors (Loughry et al. 1998a, 1998b).

BOX 6.2   Inclusive Genetic Fitness and Kin Selection

In 1964, the famous evolutionary biologist William Hamilton formally introduced the important concept that an individual in any species can transmit copies of its genes to the next generation in two ways: directly (by producing offspring) or indirectly (by helping close relatives leave descendants). The former is *personal* genetic fitness, whereas the latter contributes to an individual's *inclusive* genetic fitness. Hamilton demonstrated mathematically that any gene-encoding nepotistic behavior can in principle spread in a population if the cost that it imposes on an individual's reproduction is more than offset by the enhanced transmission of copies of that gene to the nepotist's relatives. Such kin selection should operate far more effectively on close relatives (who likely share copies of a specified gene) than on distant cousins (who likely do not). Clonemates (such as identical twins) are obviously the nearest of genetic kin. Thus, species that routinely display polyembryonic pregnancies might be especially prone to kin-selection pressures that favor the evolution of nepotistic behaviors.

Before Hamilton (1964), genetic fitness had been defined as an individual's (or a genotype's) average reproductive success in comparison to that of other individuals (or other genotypes) in the population. Inclusive fitness entails a broader view of genotypic transmission across the generations because it incorporates both the individual's personal or classic fitness and the probability that the individual's genotype is passed on by genetic relatives. These latter transmission probabilities in turn depend on how close the kin are. Concepts of inclusive fitness were advanced to explain the evolution of "self-sacrificial" behaviors, wherein any alleles for altruism may have spread in populations under the influences of kin selection. According to "Hamilton's rule," an allele will tend to increase in frequency if the cost C it entails (loss in personal fitness through self-sacrificial behavior) relative to the benefit B it receives via increased reproduction of kin is less than r (i.e., whenever $C/B < r$). Thus, in one proverbial example of "altruism," individuals could improve their inclusive genetic

(continued)

BOX 6.2 (*continued*)

fitness by sacrificing their own life for those of two full sibs, four half-sibs, or eight first cousins!

Although the twin topics of kin selection and inclusive fitness remain controversial in some evolutionary circles (Nowak et al., 2010; Nowak and Highfield, 2011), unquestionably they have added exciting new perspectives to traditional Darwinian reasoning.

An entirely different explanation of constitutive polyembryony in armadillos appears more plausible (Galbreath 1985). At the outset of her pregnancy, a female armadillo's uterus is highly constricted and has only one implantation site. Only later does the womb enlarge and make room for additional offspring, who then arise polyembryonically. This situation contrasts with the physical setup in many other mammalian species, in which multiple implantation sites exist and more of the female's uterus can thereby serve to initiate a multipup pregnancy. Thus, under the "uterine-constraint hypothesis" (Prodöhl et al. 1998; Loughry et al. 1998b), obligate polyembryony in armadillos might be a highly evolved strategy that has been maintained by natural selection because it ameliorates the severe limit on litter size otherwise imposed by the peculiar physical configuration of the maternal uterus. Thus, from the fitness perspectives of both mother and offspring, polyembryony in armadillos could be interpreted as "making the best of the available anatomical situation."

Although the uterine-constraint hypothesis for constitutive polyembryony in armadillos remains speculative, it does illustrate how an enigmatic biological phenomenon might have an unanticipated adaptive explanation after all. However, even if the uterine-constraint hypothesis is true, it leaves unanswered the deeper question of why armadillos evolved a constricted uterus in the first place.

Dizygotic Twinning

Marmosets and tamarins (fig. 6.4) are small arboreal primates that normally give birth to two or more fraternal (nonidentical) offspring per pregnancy. These species live in highly social groups and are atypical among primates in the degree to which one female in each troop monopolizes reproduction (Anzenberger 1992). Furthermore, adult and subadult nonbreeders (who typically are progeny from the breeding couple's earlier litters) routinely help their parents rear additional offspring.

FIGURE 6.4   Golden lion tamarin, *Leontopithecus rosalia* (Callitrichidae), a primate that consistently gives birth to dizygotic twins.

Factoid: Did you know? Marmosets and tamarins (family Callitrichidae) are among the world's smallest primates, and adults in some species reach only 5 inches in length (not counting the equally long tail).

Why would nonbreeding individuals forego personal reproduction to help others rear offspring? Most researchers interpret such "alloparental care" in any species as being potentially adaptive for the helpers for one or both of the following reasons. First, substantial but indirect genetic benefits might accrue to any caregiver who is the son or daughter of the breeders because some genes that the helper shares with its parents would be transmitted to new brothers and sisters. Thus, by becoming a helper, a genetically related caregiver might improve its "inclusive genetic fitness" via kin selection (box 6.2). Second, for related and unrelated caregivers alike, reproductive assistance could be self-serving if the helper learns valuable social or parenting skills that it might later utilize if it gets an opportunity to breed. Unfortunately, little empirical evidence exists either to support or refute such hypotheses for marmosets and tamarins in nature.

An astonishing discovery, based in part on molecular markers, is that each individual marmoset or tamarin is a genetic chimera (Benirschke et al. 1962): a creature whose somatic cells collectively are a mixture of distinct genotypes (Signer et al. 2000). This peculiar genetic outcome seems to be almost unique to marmosets and tamarins, where it happens routinely as follows. These primate twins start life as separate fertilized eggs (each with its own distinctive diploid genotype), but during the first month of pregnancy the embryos partially fuse inside their mother's uterus, exchanging blood cells and some other body tissues. Although the fraternal twins become physically separated before birth, each member of a set of twins in effect is genetically partly itself and partly its brother or sister.

Discovery of this odd state of affairs prompted the development of a sophisticated mathematical argument for the evolution of alloparental care in these primates (Haig 1999b). The hypothesis hinges on the suspicion that by virtue of being a chimera, each tamarin is related more closely on average to its parents than to its own sex cells (sperm or eggs). If so, then one of the two sets of genes in each chimeric twin would, in effect, devalue the animal's personal reproductive efforts in relation to any behavior directed toward helping its parents. Thus, genetic chimerism itself tips the scales toward the evolution of cooperative behavior in marmosets and tamarins by predisposing each chimeric animal to suppress personal reproduction in favor of helping its parents. As detailed by Haig (1999b), placental chimerism might also have several evolutionary implications for kin interactions within the womb, including how natural selection could promote both antagonistic and cooperative interactions between sibling embryos. Haig's intriguing genetic hypothesis remains to be tested empirically, but it does highlight the broader evolutionary notion that the mammalian womb is a key biological arena wherein selective interactions between sibling genotypes (as well as between parental and offspring genotypes) are likely to transpire. These genetic hypotheses also highlight the point that natural selection and adaptation are not synonymous; natural selection is a relational process involving differential reproduction among genotypes, whereas adaptation is some absolute measure of how well an organism matches environmental demands. In the case of marmosets and tamarins, dizygotic twinning can clearly influence the way in which natural selection operates during (as well as before and after) a pregnancy, but the phenomenon itself need not be interpreted as an evolutionary adaptation. Instead, intrauterine chimerism might just be a coincidental biological condition (or perhaps a phylogenetic legacy) with selective consequences.

Cooperation and Conflict Within the Nuclear Family

Pregnancy might seem to be the ultimate collaborative endeavor between individuals because (a) a mother and her fetus both have a vested personal interest in a successful outcome and (b) so too does the father. Indeed, all three participants (sire, dam, and fetus) in a pregnancy would seem to have a coincident mutuality of interest that progeny are born healthy after a productive incubation. On the other hand, each female mammal alone bears the physical burden of incubation and nursing, whereas the male may have little or no reproductive involvement beyond his original genetic contribution. Also, in nearly all sexual species each family member has a unique genotype, implying that a gene's optimal tactic for self-perpetuation depends at least to some degree on which of the individuals house that gene and any of its copies. Furthermore, the selfish genetic interests of interacting organisms tend to be aligned only insofar as those individuals are related (Hamilton 1964; Mock and Parker 1997; Clutton-Brock 1991), and coefficients of genetic relatedness (r) between pairs of individuals in a nuclear family vary widely. For example, a mother and her offspring normally share half their genes ($r=0.5$), as do full sibs in a multibirth litter; however, half-sib progeny share only one-quarter of their genes ($r=0.25$), and a sire and dam typically are unrelated ($r=0.0$).

For these and other reasons, each nuclear mammalian family is not simply a serene setting for harmonious interactions but can also be an evolutionary minefield of oft-competing genetic fitness interests, both inter- and intragenerational. Furthermore, many of these conflicts play out forcefully during each pregnancy (i.e. within the mammalian womb) (box 6.3). Of course, maternal-offspring relations entail many elements of cooperation as well as conflict; these two categories of interaction need not always be interpreted as mutually exclusive (Strassmann et al. 2011).

BOX 6.3 Trench Warfare at the Maternal-Fetal Interface

Haig (2010a) recently reviewed the history of scientific thought about physiological happenings at the anatomical boundary between a pregnant mother and a fetus. Before the 1890s, many physicians viewed implantation as basically a cooperative process, in which the mother's placenta provided a spongy bed that welcomed and nourished the nascent embryo. Another metaphor conveying this peaceful sentiment was that the lining, or *decidua*, of the mother's uterus acted like fertile soil that supported the growth of the embryo as the latter developed within the womb. However, embryological discoveries in the late 1890s soon changed such amicable metaphors to alternative descriptions of implantation as a hostile process in which the embryo was portrayed as a ruthless invader against which maternal defenses were deployed at the decidual interface.

Indeed, some biologists soon began to apply explicit battlefield imagery to the implantation process and to other maternal-fetal interactions during a pregnancy. For example, in describing the maternal-fetal placental boundary, Johnstone (1914, 258) wrote the following: "The border zone . . . is not a sharp line, for it is in truth the fighting line where the conflict between maternal cells and the invading trophoderm [the part of the blastocyst that attaches to the uterine wall] takes place, and it is strewn with such of the dead on both sides as have not already been carried off the field" (quoted in Haig 1993, 500). Similarly, Page (1939, 292) described the placenta as a "ruthless parasitic organ existing solely for the maintenance and protection of the fetus, perhaps too often to the disregard of the maternal organism" (also quoted in Haig, 1993, 516). Metaphors involving parasitic associations, trench warfare, and other battleground imagery were sometimes extended to other aspects of mammalian pregnancy as well. Such metaphors were motivated not by any philosophical or theoretical argument but were meant simply to summarize empirical observations made under the microscope about cellular happenings at the maternal-embryo interface. Another half-century would pass before an evolutionary rationale would be advanced that helped to make sense of the physiological, cellular, and biochemical turmoil between a mother and her fetus.

That breakthrough came in 1974, when evolutionary biologist Robert Trivers published a revolutionary manuscript titled "parent-offspring conflict." In that work, Trivers extended Hamilton's (1964) kin-selection logic by verbally and graphically formalizing the notion that conflict (rather than pure cooperation) is an inherent evolutionary feature of interactions between parents and their progeny in sexually reproducing species. Taking into account an offspring's inclusive fitness (rather than solely the

(continued)

BOX 6.3 (*continued*)

conventional perspective of a parent's reproductive success), Trivers surmised that parents and their progeny should routinely disagree about the magnitude and duration of parental care, assuming that natural selection has shaped these actors' behavioral proclivities. Furthermore, various expressions of this conflict are to be expected throughout the period of parental care, which in female mammals normally reaches its zenith during pregnancy (i.e., in the interval between implantation and parturition). (However, parental care in many mammals in effect begins even before conception [as in den-building species such as beavers], and it can extend well beyond birth and weaning, as in pack-hunting species such as wolves). Nevertheless, parturition marks an important biological transition because it represents essentially the last good opportunity for fetal genes to use biochemical and physiological mechanisms to manipulate maternal responses directly. Although genes in progeny may remain under selection to manipulate parents long after parturition, their modes of action then are likely to be behavioral as opposed to biochemical.

Evolutionary Conflict in Mother-Offspring Relations

In the early 1970s, Robert Trivers (1972, 1974) introduced an evolutionary interpretation of parent-offspring conflict. Trivers's basic insight was that sexually produced offspring are typically under selection to demand more resources from their mothers than the latter will have been selected to provide. Furthermore, the same argument applies to fathers, with the added stipulation that a male might be expected to resist an offspring's demands even more strenuously than the mother, especially if he is uncertain about genetic paternity. When viewed in this evolutionary manner, mammalian pregnancy becomes a prime biological arena for intergenerational conflict over parental resources: each offspring is under selection to seek as many parental resources as possible, limited only by any negative effects on its inclusive fitness (see box 6.2) that such demands impose on copies of its genes carried by littermates or parents (all of whose reproductive successes might be diminished if they acquiesced fully to the focal offspring's demands). Nearly all mammals display female pregnancy and female lactation, so many of these inherent parent-offspring conflicts boil down to disputes between a dam and her progeny. The net result of each such evolutionary "tug-of-war" (Moore and Haig 1991) between mother and child is some ontogenetic balance in which each offspring must settle for fewer maternal resources than it might ideally wish, and a mother surrenders more resources than she might otherwise prefer to offer. However, by evolutionary reckoning, any such maternal-fetal compromise during or after a pregnancy is less the result of a harmonious mutualism than it is an outcome of conflict mediation (see box 6.3). Indeed, if this were not true, mammalian pregnancy would probably have evolved to be less traumatic for the mother and less dangerous for her gestating embryo(s).

In his formulation of parent-offspring conflict, Trivers (1974) defined parental investment as any investment by the parent in an individual offspring that affords fitness benefits (B) to that offspring at a cost (C) to the parent's ability to invest in *other progeny*. Thus, by definition, a fitness trade-off attends any such parental expenditure. From a parent's perspective, further investment in the focal offspring is favored if $B > C$; however, from an offspring's perspective, further investment is favored when $B > rC$, where r is the chance that another sibling shares with the focal individual a copy of a randomly chosen gene. Thus, when $r \leq 0.5$ (as is true for littermates in almost all sexual species), genes in each offspring are normally under selection to seek more resources from the pregnant parent than genes in the latter are under selection to supply.

Trivers (1974) applied his ideas mostly to postpartum parent-offspring conflict, but the theory similarly pertains to maternal-fetal interactions within the mammalian womb, as detailed in an important series of papers by David Haig (1996a, 1996b, 1996c, 1999a, 2004a, 2004b, 2007). In particular, Haig's (1993) paper, titled "genetic conflicts in human pregnancy," extended Trivers's evolutionary reasoning by elaborating many physiological skirmishes that arise routinely during mammalian pregnancy. Consider, for example, fetal growth and nutrition. Early in each pregnancy, a mother stores lipids in preparation for later gestation and lactation, but her fat supplies have peaked and begun to decline by the end of the second trimester as fetal growth diminishes maternal reserves and stresses both mother and child, especially if the dam's food intake becomes limited. Perhaps partly for this reason, each pregnant female appears to impose a physiological upper limit on how much nutrition she delivers to her progeny in utero. Maternal-fetal conflicts likewise arise and play out with respect to calcium metabolism (Haig 2004a), uteroplacental blood flow (Haig 2007), placental hormones (Haig 1996c), and a host of other metabolites and metabolic signals between mother and child (Wilkins and Haig 2003).

To describe the many maternal-fetal conflicts during a pregnancy, Neese and Williams (1994, 197) stated the situation as follows: "From the mother's point of view, benefits given to the fetus help only half of her genes, so that her optimum donation to the fetus is lower than the amount that is optimal for the fetus. She is also vulnerable to injury or death from the birth of too large a baby. The fitness interests of the fetus and the mother are therefore not identical, and we can predict that the fetus will have mechanisms to manipulate the mother to provide more nutrition and that the mother will have mechanisms to resist this manipulation." The authors then went on to suggest that serious medical problems commonly arise when the balance of power between mother and fetus becomes disrupted for any reason.

Common Medical Problems During Pregnancy

One example of a serious medical condition in many human pregnancies involves a substance known as human placental lactogen (hPL), which ties up maternal insulin such that maternal blood-glucose levels rise and the fetus thereby receives extra glucose (i.e., more nutrition). In response to this biochemical manipulation by the fetus, the mother in effect develops insulin resistance and becomes predisposed to secrete more insulin, which in turn stimulates the fetus to secrete more hPL (Neese and Williams 1994). If the mother also happens for any genetic or other reason to be deficient in her physiological capacity to produce insulin, the net result can be gestational diabetes and hyperglycemia (too much glucose in the blood), which, in the extreme, may be fatal to both mother and child. Although this latter outcome is clearly undesired by both parties, it can be interpreted as an unhappy evolutionary byproduct of the greedy behavior of an offspring in its hegemonic attempts to extract extra maternal resources, in this case using hormones as a means of waging physiological warfare while still in the womb.

Another example of maternal-fetal conflict involves preeclampsia (high blood pressure), a second major scourge in human pregnancy (Redman 1989; Haig 1993; Neese and Williams 1994). Preeclampsia begins in the early stages of a pregnancy when placental cells destroy uterine nerve fibers and muscles that otherwise adjust the blood flow from the mother to the placenta. The result is a cascade of physiological effects that may include a constriction of arteries in the mother, an increase in her blood pressure (hypertension), delivery of more blood to the placenta, and a greater transfer of nutrients to the fetus. Under an evolutionary interpretation, selfish conceptuses across the generations have experienced selection pressures that have promoted

---

Factoid: Did you know? Gestational diabetes occurs in about 4% of human pregnancies, meaning that approximately 135,000 pregnant women per year experience this condition in the United States alone.

Factoid: Did you know? Preeclampsia affects about 5% of human pregnancies and is a leading cause of perinatal mortality, preterm birth, and maternal morbidity (Aubuchon et al. 2011).

such mechanisms to garner maternal resources, even if by doing so they sometimes compromise their mother's health and may even threaten her life.

A third possible example of maternal-fetal conflict involves human chorionic gonadotropin (hGH), a hormone produced by the conceptus that enters the mother's bloodstream and impels her body to retain (rather than abort) the fetus (Neese and Williams 1994). Without such fetal manipulation of maternal physiology, mothers otherwise seem to have evolved surveillance mechanisms to recognize and reject defective fetuses. Such maternal screening mechanisms make evolutionary sense because a female can increase her lifelong genetic fitness by detecting and aborting any defective conceptus early in a pregnancy (and then perhaps starting over), even if this comes at the increased risk of occasionally killing a normal embryo. By secreting high levels of hGH, the fetus in effect is trying to prove its good health to a mother who has evolved physiological mechanisms that make her suspicious of such biochemical manipulations by the embryo.

Sometimes maternal-fetal gamesmanship during evolution can play out in ways that are subtler or far less disastrous for a pregnancy than are gestational diabetes, preeclampsia, or spontaneous abortion. An interesting case in point is the common phenomenon of "morning sickness" or "pregnancy sickness," wherein expectant mothers become easily nauseated and show strong aversions to or cravings for specific foods. One proximate explanation is that high levels of hGH contribute to this sensitivity, in which case morning sickness might be viewed as an unhappy evolutionary by-product of the genetic conflict between mother and embryo (Haig 1993; Forbes 2002). Another evolutionary interpretation is perhaps more conventional (Hook 1978): that mothers have evolved physiologies that confer sensitivities to any environmental

Factoid: Did you know? Pregnancy sickness in its most extreme forms (known as *hyperemesis gravidarum*) can be fatal due to electrolyte imbalances from persistent nausea and vomiting, which may extend well beyond the first trimester of a pregnancy.

contaminants that might compromise an otherwise successful pregnancy (Profet 1988, 1992; Sherman and Flaxman 2001). According to this view, morning sickness could be part of an adaptive response that discourages the ingestion of toxins and promotes the ingestion of healthy foods during a pregnancy. Under either explanation, pregnancy sickness is not an inexplicable pathological condition but a syndrome that emerged in response to selective pressures in the stressful biological environments in which humans evolved.

Immunological Challenges

Immunological rejection of an embryo by its mother is another potential source of parent-offspring conflict within the womb, especially in mammals with refined histocompatibility systems that otherwise recognize and attack foreign tissues (Medawar 1953). Because a sexually produced fetus differs genetically from both of its parents, why does the mother's immune system not perceive the conceptus as alien and attempt to destroy it (as it would do for an invasive pathogen or a parasite)? The general answer is that viviparous mammals have evolved mechanisms that circumvent rejection responses during a pregnancy and thereby permit the dam to gestate "nonself" tissue. These physiological arrangements are not necessarily ideal from a health perspective, but they do constitute a workable accommodation between mother and fetus that normally permits the pair to survive the protracted intimate contact that mammalian pregnancy entails. The immunosuppressive mechanisms that operate during a pregnancy are especially remarkable because, throughout the entire evolutionary process by which they arose, both mother and fetus had to remain able to fend off infections by bona-fide parasites and pathogens. Thus, immunosuppression could not be accomplished simply by shutting down the immune systems of parent and child.

Instead, eutherian mammals appear to have evolved several other routes to fetal survival in the face of maternal immunosurveillance. Three of the primary physiological pathways are as follows: (a) fetal modification of the expression of transplantation antigens in the conceptus; (b) fetal alteration of the maternal immune system to its own advantage; and (c) placental limitations on the passage of effector molecules that otherwise might implement the immune response. All of these categories of immunological circumvention have been documented and illuminated to varying degrees in various mammals. However, the mechanistic details are complicated (box 6.4), and they also seem to vary considerably among species, for reasons that remain poorly understood (Bainbridge 2000). One possibility is that the highly variable lengths of gestation in different mammals produce different selective pressures that have yielded diverse evolutionary responses to the immunological challenges of pregnancy. Or perhaps much of the interspecific variation is due to historical contingencies and phylogenetic legacies. Only further research will tell. Suffice it here to say that any such physiological mechanisms of immunosuppression during a mammalian pregnancy have clearly evolved to minimize the risk that a mother will treat her fetus as a deleterious invasive parasite.

BOX 6.4   The Major Histocompatibility Complex (MHC)

As an example of the biochemical complexity underlying immunological interactions between mother and fetus, consider the expression of a set of loci in the "major histocompatibility complex" (MHC), whose products are responsible for most rejection responses in tissue transplantations. Of special importance are "class I" MHC loci, which typically have two major types of interaction with immune cells: (a) activation of cytotoxic T lymphocytes (white blood cells) if the latter bear receptors that recognize a specific MHC-peptide combination in the foreign cell; and (b) inhibition of attack by natural killer cells that bear other combinations of MHC receptors. The net effect of these mechanistic operations is to enable an individual to

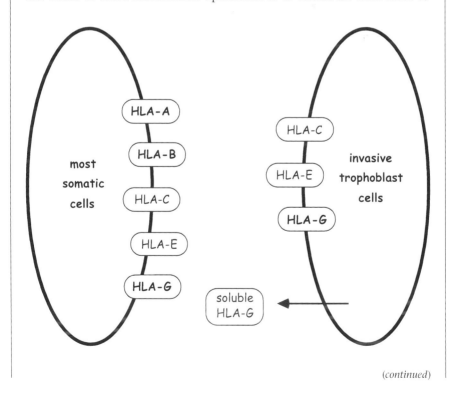

(continued)

BOX 6.4   (*continued*)

accept or reject specific foreign or domestic cells depending on how their MHC products are expressed. Accordingly, much scientific interest has centered on MHC gene expression in tissues of the placental region and in the trophoblast (the nonembryonic portion of the blastocyst that attaches to the uterine wall and later develops into the fetal portion of the placenta). For example, one basic immunological question has been whether cells of the placenta evade rejection by reducing their expression of molecules that otherwise would provoke the immune response.

   Interestingly, the answers have not always proved to be straightforward or easy to interpret. For example, although a reduction in the expression of MHC genes (especially those of paternal origin) has been demonstrated in the placental tissues of some mammals (Antczak 1989), a puzzling finding is that MHC expression is often reestablished in trophoblast cells (perhaps because this somehow either aids the invasion process or exerts an immunoprotective effect on the fetus; Bainbridge 2000). In any event, it seems to be true (as shown in the diagram) that trophoblast cells express a suite of MHC products that differ from most body cells, and immunologists are still exploring many ramifications of these differences for a pregnancy.

Infanticide

In the extreme, a pregnant female who is under exceptional physiological stress may "decide" to resorb some or all of her current litter and perhaps start over when environmental conditions improve. This practice of "brood reduction" is known to occur routinely in sheep, rats, and rodents, for example (Hayssen 1984). Although maternally initiated abortions of this sort might seem shocking, any maternal genes that encode such behaviors are likely to be selectively favored when resorption preferentially involves defective embryos or when the mother's future reproductive success might otherwise be compromised (Haig 1990). By evolutionary logic, we might predict that most such spontaneous abortions occur very early in a pregnancy (as indeed they do), before mother and fetus become intimately connected via the placenta and when the female is still young and perhaps healthy enough to start a pregnancy anew if need be. We might also predict that in response to the possibility of resorption by their dams, early-term embryos will have evolved hormonal or other countermeasures (such as secreting hGH, as described earlier) to avoid this form of infanticide.

   Another type of infant mortality occurs when an adult (sometimes even a parent) kills postpartum young. Such infanticide occurs occasionally in

> **Factoid: Did you know?** Group-living female mongooses have evolved synchronized pregnancies, and recent research (Hodge et al. 2011) suggests an explanation based on selection against outlier pregnancies. Namely, offspring born too early suffer high rates of infanticide by adults, whereas infants born too late suffer intense competition from other juveniles.

mammals and many other taxa (Hausfater and Hrdy 1984; Parmigiani and Vom Saal 1994), and it appears to have different biological motivations, depending on the species and circumstance. For example, in some cases an adult might kill even its own offspring when both are directly competing for a limited resource such as food, water, territory, or a mate. For an adult male, such shocking behavior can nevertheless make some evolutionary sense when he has a clear physical superiority and especially when he has a poor assurance of paternity for the offspring he attacks (as might be expected, for example, in species that are highly polygamous because the nonrelated victim would unlikely carry copies of the murderer's genes). Indeed, when a male almost surely is not the biological sire of the infants he assaults, infanticide can sometimes pay double dividends. For example, when an outsider lion takes over a pride, he often kills the kittens and thereby not only destroys the genes of competitor males but also brings each resident lioness into estrus and thereby sets the stage to sire future litters of his own.

Menstruation

A woman's monthly bleeding might itself be interpreted as an evolutionary outcome of escalating maternal-fetal conflict (Haig 2010a). Implantation has sometimes been likened to a maternal interview of a conceptus prior to the mother's full commitment to a laborious pregnancy. In ancestral primates from which humans descended, conceptuses rejected very early in a pregnancy may have differentially involved those that were only superficially (rather than deeply) implanted in their mother's uterine lining. At the end of each estrus cycle, the flushed material might thus have been little more than the outermost layer of endometrium, and therefore this light menstruation would selectively have favored the survival of any embryos that had invaded the interstitial uterine lining more aggressively. Thus, across the generations, embryos would presumably have evolved tendencies to invade the endometrium ever more deeply, and, in response, females would have evolved stronger tendencies to shed the endometrium in order to maintain an effective screen for potentially defective applicants to the womb. Thus, for the primate lineage leading to humans, an evolutionary hypothesis worthy of further consideration is that "a history of facultative abortion, embryonic evasion of quality control, and maternal responses may help to explain the origin of interstitial implantation and copious menstruation" (Haig 2010a, 167).

Other Categories of Conflict

Maternal-fetal disputes are not the only types of evolutionary conflict that mammalian pregnancy promotes. The following sections describe two other categories of biological warfare (among siblings and between parents) that arise or tend to be exacerbated in viviparous and other species that practice extensive parental care of offspring.

Sibling Competition

Most human pregnancies entail singletons, but multibirth pregnancies are the norm in many other mammals. Whenever two or more siblings share a womb, opportunities arise for prebirth intersibling competition for maternal resources such as space and nutrition. Sometimes this struggle can escalate into extreme outcomes such as in utero siblicide (the murder of one sibling by another while both are still in the womb). In mammals, this phenomenon has been well documented in the pronghorn, *Antilocarpa americana* (fig. 6.5). In this ungulate species, the first blastocyst that implants in a mother's uterus routinely sends out invasive tissues that pierce and kill some of the other embryos (O'Gara 1969). One net result of such in utero siblicide is that a female pronghorn normally gives birth to twins at most despite the fact that she may have ovulated several eggs that were fertilized in quick succession during her reproductive cycle.

In many other mammals, a less overt form of siblicide in effect occurs postpartum, when suckling by a newborn instigates hormonal changes in the mother's body that may block a subsequent pregnancy's initial stages (such as the implantation or preliminary development of another conceptus). Indeed, early embryonic death has been documented as an extended physiological consequence of lactation in several species of domestic ungulates (Jainudeen and Hafez 1980). More generally, however, the incidence rate of such preimplantation siblicide is hard to quantify, partly because a female is likely to resorb remnants of any aborted embryo long before her pregnancy becomes recognizable to researchers and partly because it is unclear in specific instances whether the embryonic deaths should be considered murder by a sibling or infanticide by the mother. By contrast, postimplantation siblicide is somewhat easier to identify and has been reported in several mammalian species (Hayssen 1984). Especially in mammals in which a mother delivers extensive postpartum care to her offspring via lactation, any pre- or

FIGURE 6.5   The pronghorn, *Antilocarpa americana*, a North American ungulate mammal with frequent in utero siblicide.

postimplantation siblicide promoted by suckling can be interpreted as selfish endeavor by the newborn to short-circuit subsequent pregnancies and thereby promote undivided attention from its mother. Nonetheless, as with all topics reproductive, there is a limit to how far selection can go in promoting such self-serving behavior because the mother, too, has her own vested interests in renewed reproduction, as do any genes in her earlier offspring, which would have copies represented in any additional full-sib or half-sib kin that the focal female might bear.

Sibling rivalries and competition for parental resources can extend long beyond infancy as offspring continue to vie for their parents' finite resources (Hudson and Trillmich 2008). In extreme cases, these conflicts can escalate even to the point of postpartum siblicide, which is common in many nestling birds, for example. Finally, even in species such as humans, which give birth mostly to singletons (i.e., are monotocous), siblings from successive pregnancies are likely to compete for parental resources and may attempt to manipulate their parents (behaviorally, psychologically, or otherwise) even

> **Factoid: Did you know?** Nipple stimulation seems to be the proximate factor responsible for the reproductive inhibition and long interbirth intervals experienced by mothers in some primate species (Gomendio 1990).

long after their own parturitions (Clutton-Brock 1991; Trillmich and Wolf 2007). Sometimes such attempts at parental manipulation might seem subtle (Sikes and Ylönen 1998), as in kangaroos, where lactation results in the deferred implantation (diapause) of a waiting embryo (Tyndale-Biscoe and Renfree 1987), or in many primates where lactation and nursing temporarily suppress a mother's fertility (Simpson et al. 1981).

Parental Conflicts and Genetic Imprinting

Selective pressures that pregnancy promotes have led to some outcomes that were entirely unanticipated by the scientific community. One such phenomenon is *genetic imprinting*: the expression of a gene that is inherited from one parent but not from the other (Solter 1988). In such cases a gene can have very different effects on offspring (and therefore on the course of a pregnancy), depending on whether it was transmitted via the dam (egg) or sire (sperm). Genetic imprinting in animals appears to be confined mostly to viviparous mammals, but the phenomenon is also common in plants (Scott and Spielman 2006; Feil and Berger 2007; Suzuki et al. 2007). In recent years, scientists have discovered imprinted genes in many marsupial and placental mammals, including *Homo sapiens*, where imprinting has been documented at approximately 100 loci to date. Mechanistically, imprinting usually results from the addition of methyl ($-CH_3$) groups to particular nucleotides during the production of male or female gametes, resulting in the specific inactivation of either maternal or paternal genes in offspring (Reik and Walter 2001). The terms *padumnal* and *madumnal* refer to paternally and maternally derived alleles in offspring (in contradistinction to paternal and maternal alleles present in sires and dams, respectively). Thus, genetic imprinting essentially involves altered expressions of madumnal or padumnal alleles (Haig 1996c).

> **Factoid: Did you know?** Imprinted genes account for only about 0.1–0.5% of a mammalian genome, but they have a disproportionately large influence on embryonic growth and development.

Haig (1993, 495) introduced evolutionary interpretations of genetic imprinting (and of various other expressions of conflict during mammalian pregnancy):

"The effects of natural selection on genes expressed in fetuses may be opposed by the effects of natural selection on genes expressed in mothers. In this sense, a genetic conflict can be said to exist between maternal and fetal genes. Fetal genes will be selected to increase the transfer of nutrients to their fetus, and maternal genes will be selected to limit transfers in excess of some maternal optimum. Thus a process of evolutionary escalation is predicted in which fetal actions are opposed by maternal countermeasures. The phenomenon of genomic imprinting means that a similar conflict exists within fetal cells between genes that are expressed when maternally derived, and genes that are expressed when paternally derived."

In other words, in addition to standard parent-offspring and offspring-offspring conflicts during a pregnancy, we might expect to observe what in effect are parent-parent conflicts with respect to how madumnal versus padumnal genes impinge on resource-acquisition behaviors (and associated growth profiles) of embryos within the mammalian womb. Haig's seminal ideas have become known as the "conflict hypothesis" or the kinship hypothesis for genetic imprinting. Although other theories of genetic imprinting have been advanced (box 6.5), the conflict hypothesis remains the leading evolutionary explanation for the imprinting phenomenon.

BOX 6.5   Alternatives and Extensions to the Conflict Theory of Genomic Imprinting

At least two other proposals exist for the evolution of genomic imprinting (Haig 2003). First, McGowan and Martin (1997) and Beaudet and Jiang (2002) proposed that imprinting evolved because it imbued populations with increased evolvability. Their basic idea is that whenever an imprinted allele is silenced for one or more generations and thereby effectively masked from exposure to natural selection, it becomes freer to accumulate mutations and thereby adds to the pool of hidden genetic variation that sooner or later might contribute to evolutionary responses to environmental change. This hypothesis at face value is unappealing to many evolutionary biologists not only because it seems to invoke group selection but also because it is vague about which particular genes should be imprinted and why those genes are marked but not others. Second, Varmuza and Mann (1994) proposed that oocyte genes responsible for trophoblast formation

(continued)

BOX 6.5   *(continued)*

(and perhaps some "bystander" loci also in the oocyte) are inactivated so as to prevent diseases (such as teratomas) that can otherwise arise when an unfertilized egg implants inappropriately. Formal evolutionary models have shown that this "ovarian time bomb" hypothesis is at least plausible (Weisstein et al. 2002), but whether it is the correct general explanation of genomic imprinting remains to be determined.

Apart from genetic imprinting per se, several other genomic phenomena have been proposed to arise from evolutionary conflicts of interest between padumnal and madumnal genes during ontogeny. For example, Frank and Crespi (2011) suggest that such intragenomic conflict may affect the regulation of embryonic growth in ways that may precipitate various pathologies such as some cancers and psychiatric disorders (including some cases of autism and schizophrenia). More generally, these authors view evolutionary-genetic conflict as sexual antagonism that can lead to pathologies whenever opposing genetic interests that normally are precariously balanced become unbalanced for any reason. Burt and Trivers (2006) extend this evolutionary argumentation to a broad spectrum of otherwise puzzling empirical properties of genomes.

Subsequent research has shown that imprinting is most prevalent for genes expressed in tissues that are specialized for transporting maternal nutrients to the embryo, namely, the placenta in mammals and the endosperm in plants. Given the benefit of evolutionary hindsight, perhaps these two foci for imprinting are not surprising. In plants and mammals alike, mothers invest heavily in their progeny (by supplying seeds with nourishment and by becoming pregnant, respectively). Especially in such species with highly gender-asymmetric parental care, genes carried by males versus females presumably experience different selective pressures related to their prospects for survival and replication in progeny. Thus, genes transmitted to embryos from dams versus sires are likely to view the reproductive world through very different eyes, and this gender-based difference appears to play out during evolution in how specific genes become imprinted in embryo-nurturing tissues. Two such placental genes that appear to influence an embryo's acquisition of maternal nutrients during gestation are insulin-like growth factor II (IGF2) and insulin-like growth-factor-binding protein-1 (IGFBP-1), and both of these loci have been the subject of extensive investigations in an imprinting context (Tycko and Morrison 2002; Reik et al. 2003; Frost and Moore 2010; Panhuis et al. 2011). Several more such imprinted genes and their empirical effects on intrauterine growth are compiled in table 6.1.

TABLE 6.1   Examples of imprinted mammalian genes and their effects on intrauterine growth of mouse embryos[1]

Note that the paternally expressed alleles tend to stimulate embryonic growth (because knocking them out reduces fetal weight), whereas maternally expressed genes tend to depress intrauterine growth (because knocking them out boosts fetal weight).

Expression	Gene, and Its Product	Knockout Proportion of a Fetus's Normal Weight (%)
Paternal	IGF2, growth factor	50
	IGF2P0, growth factor	75
	Peg 1, α/β hydrolase	88
	Peg 3, zinc finger transcription factor	80
	Ins 1/2, insulin	80
	Slc 38a4, system A amino acid transporter	80
Maternal	H19, noncoding RNA	130
	IGF2r, IGF2 clearance receptor	140
	Ip1, cytoplasmic protein pleckstrin	100
	Grb10, adaptor protein	146
	p57^{Kip2}, cyclin-dependent kinase inhibitor	100

[1]After Fowden et al. 2006.

Factoid: Did you know? The endosperm of a plant is an extraembryonic structure that nourishes a plant embryo much as the placenta nourishes a mammalian embryo.

The basic evolutionary idea is that from a father's selfish fitness perspective, any paternal genes in his progeny should be expressed in such a way as to promote the health and survival of his offspring (each of whom shares 50% of his genes) even if this comes at the expense of the health of the mother (who normally shares 0% of her genes with the sire) and her future offspring (who might have different sires). Conversely, maternal genes should be expressed in such as way as to limit the current offspring's hegemony over maternal resources to a level that is more compatible with the mother's continued health and future reproduction. In other words, the conflict or kinship theory of genomic imprinting states that paternal genes maximize the extraction of maternal resources for the benefit of the focal sire's offspring but does so at the expense of offspring from future pregnancies, who may have different sires; by contrast, the maternal genome limits its support for each embryo irrespective of the paternal genome. Such evolutionary reasoning also highlights the fact that the phenomena of genomic imprinting and kin selection are thoroughly intertwined (Haig 2002, 2004b).

Thus, what emerges in each species during the evolutionary process is some resolution of the conflict between the optimal tactic for genes inherited from the sire versus that for genes inherited from the dam (reviews in de la Casa-Esperon and Sapienza 2003; Wilkins 2005; Haig 2000). In general, padumnal genes are under selection to promote embryos to extract the maximum possible nutrition from the mother, whereas madumnal genes tend to be under selection to relinquish fewer resources. Furthermore, many biologists now suspect that these skirmishes over maternal resources transpire precisely where they might be most expected: at the physical sites (placenta and endosperm) where mothers deploy nutrients to their embryos. Unfortunately, these strategic battles between madumnal and padumnal genes in utero come not without serious medical consequences, especially for embryos that may be caught in the evolutionary crossfire.

Consider, for example, conflicts between paternally and maternally derived genes along a small region of human chromosome 15 that is responsible for two contrasting medical disorders: the Prader-Willi and the Angelman syndromes. In normal pregnancies, paternally derived genes in this chromosomal region probably favor high infant activity (including sleeplessness and aggressive feeding), whereas maternally derived genes in the same locus have more of a restraining influence on an infant's behavior. However, when the padumnal copy of the gene is unexpressed (e.g., either because of a small deletion or via imprinting), the net result is Prader-Willi disorder, which causes the affected child to be short of stature and obese and to display poor motor skills, underdeveloped sex organs, and mental retardation. Conversely, when the madumnal copy of the gene is unexpressed (again because of a deletion or imprinting), the net result is a child with Angelman disorder, which causes slow development and certain neurological difficulties. Thus, these complementary syndromes, which originate in the same chromosomal region, exemplify at least two points relevant to the current discussion: (a) the paternal versus the maternal source of particular genes can in some cases cause large differences in offspring behavior, and (b) when the expression of the maternal or paternal copies of such genes is disrupted for any reason, interallelic conflicts may occur and sometimes have major consequences for human health.

TABLE 6.2   Human disorders related to imprinting[i]

Shown are examples of human metabolic disorders for which genomic imprinting is known to be a contributing factor in at least some cases.[ii]

Angelman syndrome. A disorder that causes delayed development and neurological problems, including jerky movements, sleep disorders, and seizures; associated with a small gene region on chromosome 15, which, when improperly imprinted or missing in progeny, results in this syndrome (see also Prader-Willi syndrome).

Fetal growth restriction. A condition in which a fetus is unable to reach its genetic potential for body size; can have many different etiologies, including errors in genetic imprinting.

Gestational diabetes. A form of diabetes (a disease in which the pancreas cannot properly produce or utilize insulin) that occurs in otherwise nondiabetic women during pregnancy.

Prader-Willi syndrome. Another serious genetic disorder associated with imprinting or other errors at a gene region on chromosome 15 (see also *Angelman syndrome*).

Preeclampsia. A disorder that affects both the mother and fetus typically in the second or third trimester; it involves rapidly progressing symptoms, including sudden weight gain, headaches, and changes in vision; affects 5–8% of all pregnancies and can be fatal.

Rett syndrome. A neurological disorder that begins in early childhood and is characterized by autistic-like behavior, language impairment, and mental retardation; caused by male germ-line mutations in an X-linked gene (MECP2) that codes for a methyl-CpG binding protein.

Spontaneous miscarriage. Termination of a pregnancy during the first 20 weeks of gestation; can have many different causes, but some cases involve imprinting errors.

Turner syndrome. A genetic disorder in girls often associated with failed ovarian development, webbed neck, drooping eyelids, heart and kidney defects, and other symptoms; caused by a missing or defective X chromosome with some of the health problems, depending upon whether the chromosome came from the sire or dam.

[i]After Avise 2010.
[ii]Many of these syndromes relate directly or indirectly to pregnancy phenomena.

Several other imprinted genes in humans are likewise known to cause serious disorders. Consider, for example, the IGF2 gene, which encodes an insulin-like growth factor. Normally, only the paternal copy of IGF2 is expressed in offspring, but if the sire's copy of IGF2 is silenced through a biochemical mishap during spermatogenesis, a child may be born with Silver-Russell syndrome, which is characterized by abnormally low birth weight and retarded growth. Conversely, if the mother's copy of IGF2 is silenced by a biochemical mishap during oogenesis, her child may display Beckwith-Wiedemann syndrome, characterized by high birth weight and symptoms of overgrowth often associated with an increased risk of tumors. As summarized in table 6.2, abnormalities in genomic imprinting similarly play leading roles in at least some clinical cases of several other common metabolic disorders during human pregnancy, including preeclampsia, miscarriage, fetal growth restriction, and gestational diabetes. Scientific evidence further suggests that some widespread psychiatric illnesses, including autism and schizophrenia, might also be linked to the imbalanced gene expression that is sometimes related to imprinting (Badcock and Crespi 2006; Crespi and Badcock 2008; Crespi et al. 2010). One scientific suggestion is that patterns of imprinted-gene expression affect not only pregnancy but also brain development (Gregg et al. 2010; Keverne 2010) and that deviations in those imprinting patterns can cause metabolic imbalances that can lead in exceptional cases to severe mental illness. If so, where one resides on a spectrum ranging from autism to normalcy to psychosis may be partly the result of one's imprinted genes, which themselves probably became imprinted ultimately because of the evolutionary "battle of the sexes" (Badcock and Crespi 2008) that mammalian pregnancy promotes.

SUMMARY

1. Pregnancy in mammals has been shaped by and in turn energizes powerful forms of natural selection, both positive and negative. However, not all pregnancy-related phenomena in viviparous mammals evidence selective effects or demand adaptive explanations; some almost certainly do, some almost certainly do not, and others are ambiguous in this regard.

2. Embryonic diapause (a delay between conception and implantation) has evolved independently in several mammalian taxa and can be highly adaptive in particular environments. Sporadic polyembryony (the occasional production of clonal siblings) occurs in several mammals, including humans, but is probably happenstantial and has no special selective significance. Both constitutive polyembryony and dizygotic twinning in particular mammalian taxa have elements that remain controversial in terms of their evolutionary motivations and their ramifications for natural selection.

3. Despite outward appearances, mammalian pregnancy is not simply a loving collaborative venture within the nuclear family (mother, father, and child); rather, it is a phenomenon rife with inherent evolutionary genetic conflicts that sometimes escalate to what has often been described as biological warfare. In pregnancy, such conflicts can be both intra- and intergenerational, and they include (a) disagreements between parents and progeny about the delivery of offspring care, (b) sibling competition both in utero and postpartum, and (c) in effect, parental disputes about how genes of maternal versus paternal origin should be expressed in conceptuses.

4. Many physiological conflicts and accommodations exist between a pregnant mother and her sexually produced fetus. Generally, each self-serving offspring is under selection to seek more maternal resources than its mother might wish to relinquish given the negative effects that such donations can have on a dam's lifetime genetic fitness. Ultimately, such conflicts are resolved as evolutionary compromises between the oft-competing genetic interests of mother and child. Proximately, maternal-fetal conflicts during a pregnancy can register as a wide array of health problems ranging from the relatively subtle (such as morning sickness) to the egregious (such as infanticide and litter reduction by spontaneous abortion).

5. One of the wonders of mammalian pregnancy is that the mother's immune system does not recognize and reject the fetus as being genetically foreign (as it would do for any other invasive parasite). Several mechanisms have evolved by which mothers and their fetuses circumvent histoincompatibility. These often fall into three categories: fetal modification of the expression of its transplantation antigens; fetal modulation of the maternal immune system to its own advantage; and placental impediments to the passage of effector molecules that otherwise implement the immune response.

6. Even menstruation might be interpreted as an evolutionary outcome of parent-offspring conflict early in a pregnancy. The process by which an embryo implants into its mother's uterus can be likened to a maternal "interview" of a conceptus prior to the dam's full commitment to a laborious pregnancy. During this interview, fetuses are under selection to avoid rejection, while the female is under selection to maintain an effective screening of womb applicants. The net result may have been increased invasiveness by the embryo and the evolution of menstruation, wherein primate females shed increasing amounts of uterine tissue in their efforts to abort defective conceptuses.

7. Especially in mammals with multipup litters, conflicts over finite maternal resources routinely arise among siblings, both in utero and postpartum. Such disputes sometimes escalate to overt siblicide, but often they are expressed in subtler ways, such as when suckling by a newborn stimulates hormonal

responses in the mother that prompt her to delay or even abort subsequent conceptions.

8. Another category of genetic conflict exaggerated by pregnancy pertains to the two parents because selfish genes of paternal versus maternal origin in progeny inevitably experience different selective pressures. In mammals, this sexual difference sometimes plays out via genetic imprinting, wherein gene expression in progeny depends on which of the two parents transmitted the imprinted gene. Understandably, genetic imprinting seems to be especially prominent in the placenta (in mammals) and the endosperm (in plants), two tissues intimately involved in supplying nutrients to embryos. Imprinted genes are now known to underlie many biochemical imbalances that cause serious metabolic disorders in humans.

CHAPTER SEVEN

Sexual Selection and Piscine Pregnancy

Sexual selection arises from the differential ability of individuals to obtain sexual partners and fertilize gametes. It is the "other form" of selection explicated by Darwin (1871). Unlike natural selection (Darwin 1859), which tends to forge adaptations in response to environmental demands, sexual selection tends to mold behaviors and other phenotypes to meet mating demands. As Darwin recognized, sexual selection is tightly interwoven with many other reproductive topics, including mating behaviors, mating systems, sex ratios, and sexual dimorphism. Even a century and a half after Darwin (Avise and Ayala 2009), modern extensions of sexual-selection theory continue to present empirical and conceptual challenges to evolutionary biologists (Eberhard 2009; Jones and Ratterman 2009; Shuster 2009). This chapter illustrates such research by focusing on sexual selection in relation to pregnancy-like phenomena in fish.

Anisogamy, Sexual Selection, and Animal Mating Behaviors

As intimated in chapter 1, anisogamy—the asymmetry in size and motility between male and female gametes—has many evolutionary implications for mating systems, modes of pregnancy, and the operation of sexual selection. These ramifications include the following (Avise 2010), all of which apply broadly (but not universally) to diverse animal taxa:

1. *greater potential fecundity (fertility) for individual males than for females.* Because sperm cells are relatively tiny and inexpensive to produce,

each male can typically produce vastly more gametes than can a female.

2. *a gender bias toward male pursuit of multiple mating.* By securing multiple mates, a male is likely to produce many additional offspring, whereas a female is not. The statistical regressions that describe empirical relationships between mate numbers and progeny output are called *sexual-selection gradients* or *Bateman gradients* (box 7.1). The slopes in these regressions are steeper for males than for females in many species (especially those with female pregnancy). Notable exceptions include some male-pregnant pipefishes (described later).

BOX 7.1   Sexual-Selection Gradients

The relative intensities of sexual selection on males and females in various species have been attributed to several influences, including differences in parental investment by the two sexes (Parker and Simmons 1996; Trivers 1972); operational sex ratio (of sexually mature males to females; Kvarnemo and Ahnesjö 1996); relative variances in reproductive success in the two genders (Payne 1979; Wade and Arnold 1980; Wade and Shuster 2004); and the two sexes' potential reproductive rates (Clutton-Brock and Parker 1992; Clutton-Brock and Vincent 1991). In a classic paper, Angus Bateman (1948) argued that all such influences can be subsumed under one overarching factor: the average relationship between an individual's mating success (number of mates) and the number of offspring the individual produces (reproductive success or genetic fitness).

In experimental populations of fruit flies, Bateman documented that males' mean genetic fitness increased rapidly as a function of mate numbers, yielding a steep sexual-selection gradient, whereas females' genetic fitness did not increase nearly as rapidly with higher mate counts, yielding a nearly flat selection gradient (see the accompanying graph). Bateman interpreted this disparity as sexual selection's true cause: multiple mating pays higher fitness dividends to males than to females.

The slopes in sexual-selection gradients (or Bateman gradients) are now routinely used to quantify and compare the relative intensities of sexual selection on males and females (Andersson and Iwasa 1996; Arnold and Duvall 1994; Jones et al. 2000, 2002; Lorch et al. 2008). Although Bateman's approach to quantifying sexual selection has several limitations (Tang-Martinez and Ryder 2005; Snyder and Gowaty 2007), many researchers now accept the basic thrust of Bateman's argument: reproductive suc-

(continued)

BOX 7.1   *(continued)*

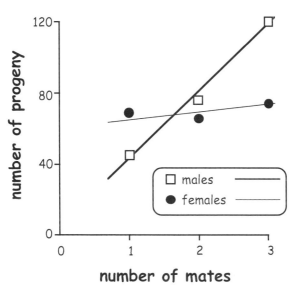

cess in males tends to be limited by access to mates, whereas in females it tends to be limited by access to resources (Jones and Ratterman 2009).

None of this is to imply that females never benefit from multiple mating. Direct benefits that might accrue to a female from having multiple partners include more nuptial gifts, improved territorial access, or additional help with parenting. A female might also receive "indirect benefits" (Jennions and Petrie 2000; Zeh and Zeh 1996, 1997, 2001; Tregenza and Wedell 2000) from multiple mating when she thereby gains higher genetic diversity in her clutch; greater opportunities for "good genes" in her progeny (Yasui 1998, 2001); "fertilization insurance" against the possibility that some males might be sterile; and/or a reduced risk of sexual harassment (Birkhead and Pizzari 2002). However, none of these factors is likely to boost a female's genetic fitness nearly as much as multiple mating by a male can normally boost his genetic fitness. Of course, multiple mating by either sex also carries costs, such as the time and energy involved in finding mates and the higher possibility of contracting a sexually transmitted disease.

3. *a bias toward the evolution of polygyny rather than polyandry. Polyandry* (multiple mating exclusively by the females of a species) is relatively rare in the biological world, whereas *polygyny* (multiple mating predominantly

by males) is common. This makes evolutionary sense given that Bateman gradients are often shallower for females than for males. However, the rarity of polyandry also has to do with asymmetrical assurances of genetic parentage. Especially in viviparous taxa such as mammals, a female can be certain that she is the biological mother of the offspring she nurtures, whereas a male has no comparable assurance of paternity. Because any male that enters into an exclusive relationship with a polyandrous female runs the risk of being cuckolded by a competitor, natural selection tends to disfavor the evolution of male acquiescence to mate with and care for a single polyandrous partner and her young.

4. *a gender bias toward the evolution of postzygotic parental care (including pregnancy) by females.* Anisogamy generally predisposes females more so than males to invest additional energy and resources in rearing offspring, a disparity that becomes exaggerated in viviparous species with female pregnancy. Thus, anisogamy biases evolution toward the emergence of gestation by females. Furthermore, female pregnancy can then amplify the following biological tendencies that anisogamy ultimately initiated.

5. *a gender bias toward higher variation in reproductive success among males.* Especially in species with polygynous mating systems, some males are huge winners in the reproductive sweepstakes, while other males may fail miserably, resulting in higher intermale than interfemale variances in reproductive output. By contrast, in any species with lifelong monogamy for all mating pairs, intermale and interfemale variances in genetic fitness are identical.

6. *a gender bias toward male competitiveness for mates.* This expectation applies especially to polygamous species, and it should also apply more strongly to males in polygynous taxa than to females in polyandrous taxa. However, this is not to imply that either gender in a monogamous species is immune to sexual selection because each individual in any mating system can presumably improve its genetic fitness by choosing a high-quality mate. Nevertheless, anisogamy (alone or especially when coupled with female pregnancy) generally makes females a limiting procreative resource, for which males compete for sexual access.

7. *a gender bias toward sexual selection on males.* This follows directly from the preceding item and again applies with special force to species with polygynous mating systems.

8. *a gender bias toward the elaboration of secondary sexual traits in males.* This follows directly from the preceding item and also applies with special force to polygynous species.

Figure 7.1 graphically summarizes these points, which collectively reflect conventional evolutionary wisdom about many standard connections between mating behaviors, sexual selection, sexual dimorphism, and mating strategies (Shuster and Wade 2003; Oliveira et al. 2008). However, none of these extended ramifications of anisogamy are universal evolutionary outcomes in the heterogeneous biological world (Bonduriansky 2009). For example, males in some species become pregnant (Vincent et al. 1992; Jones and Avise 2001), in which case many of the basic evolutionary ground rules shift dramatically. Nevertheless, each biological expectation listed earlier is met in a sufficiently wide array of taxa as to make the exceptions particularly informative. This is why the relatively few species that display male pregnancy and "sex-role reversal" are fascinating subjects for evolutionary research (see later discussion).

Categories of Sexual Selection

Intragender Versus Intergender

Modern researchers follow Darwin (1871) in dividing sexual selection into two broad categories: intersexual and intrasexual. The latter (sometimes called "male-male competition") typically favors males that are good fighters, stalwart defenders of territory, or otherwise stiff competitors against other males in their ability to secure mates. Male-male competition has resulted in the evolution of impressive traits such as huge horns on the heads of bighorn rams, fighting spurs on gamecocks, and combative behavior in Siamese fighting fish. For champion males, the ultimate fitness payoff is more offspring via enhanced mating success. By contrast, intersexual or epigamic selection (sometimes abbreviated "female choice") typically operates by favoring males that females find attractive as sexual partners. It, too, has resulted in the evolution of many impressive phenotypic features such as showy peacock tails, colorful fins on guppies, and elaborate courtship displays by males in many species, including fishes. Again, the reproductive payoff for showier males is additional progeny as a result of greater mating success.

Sexual selection of either sort can be a powerful evolutionary force that promotes striking intraspecific sexual dimorphism (consistent differences between conspecific males and females) in secondary sexual traits (gender-specific phenotypes other than the gonads). However, unless experiments or critical observations are made, it can be difficult to know in specific instances whether dimorphism in a secondary sexual trait arose via intersexual or intrasexual selection (or both). Do the showy fins of male guppies register the effects of male-male competition or female choice (or in this case perhaps

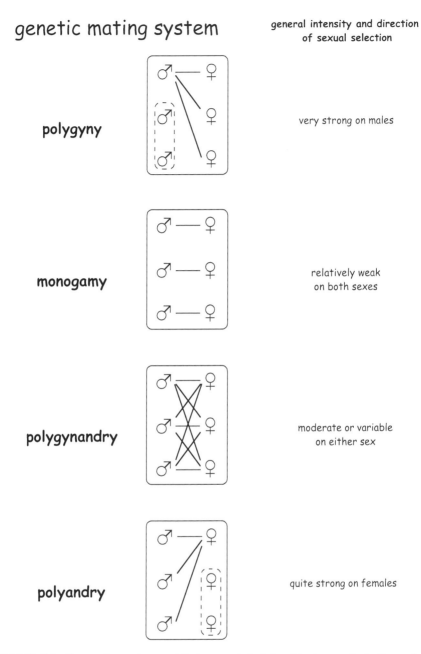

FIGURE 7.1 Four mating systems and their purported relationships to sexual selection and gender dimorphism in secondary sexual traits (modified from Avise et al. 2002). Solid lines indicate either "pair bonds" or successful mating events (depending on whether the reference is to the "social mating system" or to the "genetic mating system" of a population). Although the mating systems as depicted here are population-level phenomena, the terms can also apply to the mating behaviors of individuals. Thus, a female might be polyandrous or monogamous, depending on whether she has multiple mates or just one, and a particular male might show either polygynous or monogamous mating tendencies.

Factoid: Did you know? Many superfancy goldfish can barely swim at all, but this undoubtedly reflects extreme artificial selection by human aquarists more so than extreme sexual selection by other goldfish.

artificial selection by human aquarists)? Another point is that sexual selection and natural selection often act in opposition on a trait. Whereas a showy caudal fin may enhance a male guppy's attractiveness to females, it might also compromise his swimming ability and make him more vulnerable to starvation or predation. In each such case, evolution in essence strikes a compromise between the benefit of winning mates and the detriment of attracting predators (Endler 1991).

In principle, the genetic mating system of a population or species is pertinent to the direction and intensity of both intra- and intergender sexual selection. For example, in a population with a balanced sex ratio and a polygynous mating system, some males inevitably go without sexual partners, whereas others may obtain several mates, thus leading to a pronounced competition for females that can translate into strong sexual selection and the evolution of secondary sexual traits in males. Conversely, in a population with a 1:1 sex ratio and a polyandrous mating system, some females fail to mate, whereas others are multiple maters, thus driving sexual selection on females and promoting the evolution of secondary sexual traits in that sex. However, the disparity between the genders in the intensities of sexual selection (and in the elaboration of secondary sexual traits) is seldom expected to be as strongly female biased under polyandry as it is to be male biased under polygyny because anisogamy implies that the fecundity of females in most species remains inherently more limited than that of males. In effect, this gender-based asymmetry of fertility becomes even more pronounced in species with female pregnancy.

Female Choice Versus Male Choice

Notwithstanding the general bias toward female choosiness, active mate choice by males is also common in many animal species (Gwynne 1991; Gwynne and Simmons 1990; Gowaty and Hubbell 2009). This should come as no great surprise because individuals of both sexes surely are under selection to seek the best possible mates regardless of other circumstances. For both sexes, such a mate might be of high quality because the mate confers fitness benefits of either of two general types: better genes for the offspring or higher-quality care (more material resources) for its mate and/or for the resulting progeny. Despite this reality, it also remains true that females in most species exercise much of the authority over mate choice (chapter 1). Furthermore, especially for males, any fitness boosts obtained from acquiring a high-quality mate are likely to remain lower than those that might otherwise come from mating with multiple sexual partners (even if some of the latter are somewhat lower in quality).

Precopulatory Versus Postcopulatory Female Choice

Another way to dichotomize sexual selection is to take into account whether female "mate choice" takes place before or after copulation (Birkhead and Pizzari 2002; Birkhead 2010). For polyandrous species with internal fertilization, sperm from two or more males routinely co-occur within a female's body. Growing evidence suggests that the reproductive tract of a female often plays a more active role than previously supposed in the postcopulatory choice of fertilizing sperm (Birkhead and Møller 1992, 1993; Mack et al. 2003). Furthermore, prolonged female sperm storage (a phenomenon known to occur in many viviparous organisms, including some live-bearing fishes; Liu and Avise 2011) probably enhances the opportunity for such "cryptic female choice."

Multiple mating by a female also enhances the opportunities for postcopulatory sperm competition wherein gametes from different males undergo selection for traits (such as faster swimming speed or greater longevity within the female's reproductive tract) that might improve their odds of successfully fertilizing eggs. Many reproductive phenotypes of males and their ejaculates (including sperm motility itself) can be interpreted as adaptations to meet the genetic challenges of direct or indirect competition with other males' sperm (Smith 1984).

From a female's perspective, mechanisms to prevent competition among sperm from different males are not necessarily desirable (Keller and Reeve 1995), and this can lead to reproductive conflicts of interest between the sexes (Eberhard 1998, 2009; Knowlton and Greenwell 1984). These and related discoveries about postcopulatory phenomena have made sperm competition and cryptic female choice major research arenas in the last several decades (Parker 1970; Smith 1984; Baker and Bellis 1995; Birkhead and Møller 1992, 1998).

Female Pregnancy and Traditional Female Choice

In Darwin's era, the concept of intersexual selection was slower to gain general acceptance than was natural selection, probably because many people found it hard to believe that females in "lower" animals have the perceptive wherewithal to choose mates with special phenotypic adornments. Today,

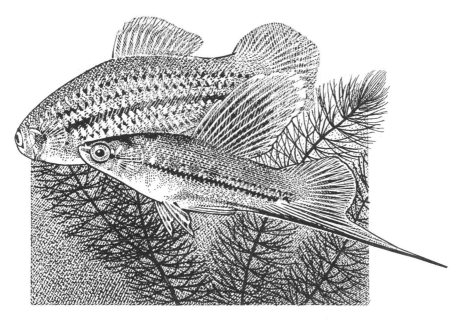

FIGURE 7.2   The swordtail fish, *Xiphophorus maculatus*, a live-bearing member of the Poeciliidae. The elongate caudal fin of the male is not an intromittent organ but rather an ornamental device that apparently evolved via sexual selection.

most researchers recognize that females in many species do indeed exercise refined choices of alternative traits in their sexual partners, and the evolutionary details of such choosiness have sometimes been worked out to varying degrees. One case in point involves female-pregnant platyfishes and swordtails (*Xiphophorus*), an ensemble of more than 20 species of popular aquarium fishes native to Central America.

With respect to body form, all platys and swordtails look quite alike, and the most obvious difference is the presence in mature male swordtails of a long caudal fin resembling a cutlass (fig. 7.2). Researchers have documented that female swordtails prefer to mate with long-tailed males, thus implying that intersexual selection was probably responsible for the evolution of this striking phenotypic feature (Basolo 1990, 1995; Endler and Basolo 1998).

Interestingly, platyfish females also prefer to mate with long-tailed males when given the option, as was shown by laboratory experiments in which researchers surgically grafted plastic prostheses onto platyfish males that otherwise lacked long tails. These observations raised a chicken-or-egg question: which came first in *Xiphophorus* evolution—male swords or female preferences for male swords? Clues to the answer came from phylogenetic

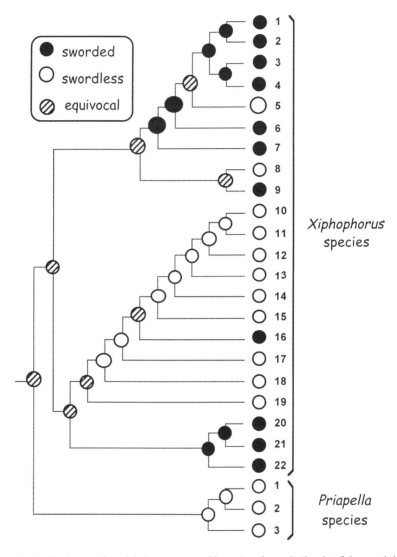

FIGURE 7.3   Evolution of caudal phenotypes in 25 species of swordtails, platyfishes, and their allies (after Schluter et al. [1997], based on the molecular phylogeny from Meyer et al. [1994]). Pie diagrams indicate the reconstructed likelihoods (from PCM analyses; see box 1.6) of alternative tail conditions at various nodes in the phylogenetic tree.

analyses. If platys were ancestral to swordtails, this would be consistent with the "preexisting-bias" hypothesis (Basolo 1995), which states that female preference for long tails came first in evolution; however, if swordtails were ancestral to platys, then swords (but not female preference for them) might later have been lost in the derived platyfish clade.

To address these issues, a molecular phylogeny for *Xiphophorus* species was generated and employed as a backdrop (fig. 7.3) for deducing evolutionary histories of the various phenotypes. This exercise in phylogenetic character mapping (PCM) revealed that swordless and swordtailed species are intermingled in the phylogenetic tree and that neither assemblage forms a coherent clade. Thus, evolutionary transitions between presence and absence of swords appear to have been both rapid and recurrent within the genus, thereby compromising attempts to determine the ancestral state of this clade (Meyer et al. 1994). Nevertheless, outlier species in the sister genus, *Priapella*, consistently lack a sword, so the shared ancestor of the more ancient *Priapella* + *Xiphophorus* clade was probably swordless. Behavioral experiments have shown that females of some *Priapella* species also prefer to mate with males adorned with long prosthetic tails. Thus, overall, the available data provide substantial evidence for the preexisting-bias hypothesis, in which a female preference for swords predated the evolutionary appearance of the swords themselves.

The phylogenetic reconstruction in figure 7.3 also indicates that swords were both lost and gained on multiple occasions within the *Xiphophorus* clade. This suggests that swordlike tails, despite their attractiveness to females, are otherwise a substantial burden for the males. Maybe the swords are energetically costly to produce, or perhaps they compromise the males' agility. Laboratory experiments have confirmed that males with long swords do indeed expend more energy during normal swimming than do males without these encumbrances (Basolo and Alcarez 2003).

Internal Male Pregnancy

In one piscine taxonomic family—Syngnathidae—males (rather than females) become pregnant. In all of the 200-plus living species of pipefishes (fig. 7.4) and seahorses (fig. 7.5), males house the developing embryos. The process begins when a gravid female transfers some or all of her many eggs to the male's abdomen or tail, where the eggs are either glued onto his external surface or deposited into a specialized pouch that evolved expressly for this purpose. In species with pouches, the male then fertilizes the clutch internally, seals the pouch, and carries the embryos for several weeks before giving birth to live young. During this pregnancy, the sire nourishes (Ripley and Foran 2009), aerates, osmoregulates (Ripley 2009), and protects his brood, whereas the mother plays no direct role in offspring care.

Evolutionary Origins

Pipefishes and seahorses are evolutionary cousins of sticklebacks (fig. 7.6), another group of fishes traditionally placed in the order Gasterosteiformes

FIGURE 7.4   A pregnant male pipefish.

Factoid: Did you know? Like a mammalian womb, a syngnathid brood pouch can be an active site for postcopulatory sexual selection and parent-offspring conflict. For example, Paczolt and Jones (2010) show that pregnant males increase abortion rates in pregnancies from unattractive mothers, presumably thereby retaining their resources for future reproductive opportunities.

(Kawahara et al. 2008). In sticklebacks, an adult male builds a globular nest in which he may care for the clutches of several females simultaneously. Thus, one possibility is that the evolution of internal male pregnancy in the syngnathid lineage entailed a shift from the ancestral stickleback type of parental care (within a nest) to the pipefish type of internal parental care (within a male), with an intermediate stage wherein a female's eggs were probably glued to the exterior of a male's body. Two other possibilities not yet excluded by available phylogenetic evidence (Wilson et al. 2003; Kawahara et al. 2008) are that the syngnathid ancestor either displayed no parental care or perhaps practiced female brooding.

Although a higher assurance of paternity via internal fertilization likely played a selective role in the evolutionary transition to an enclosed brood pouch, other selection pressures may have come into play as well. For example,

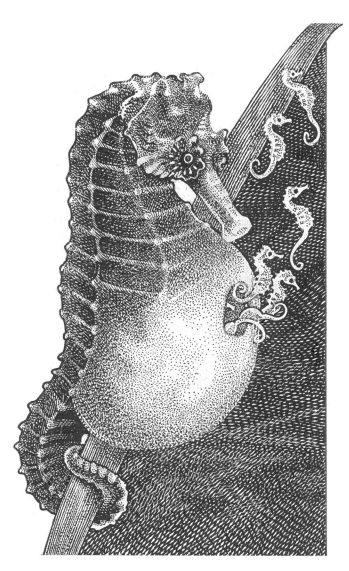

FIGURE 7.5   A pregnant male seahorse giving birth to live young.

Andersson (2005) suggests that a loss of swimming agility in an ancestral nest-tending species became exaggerated when early syngnathids evolved a slim snout for ambush hunting of small prey. As poor swimmers, these fish would have been vulnerable to predation, and in response they presumably evolved a cryptic morphology and began carrying eggs deeper within their bodies for safekeeping. These developments in turn truncated male fecundity and set

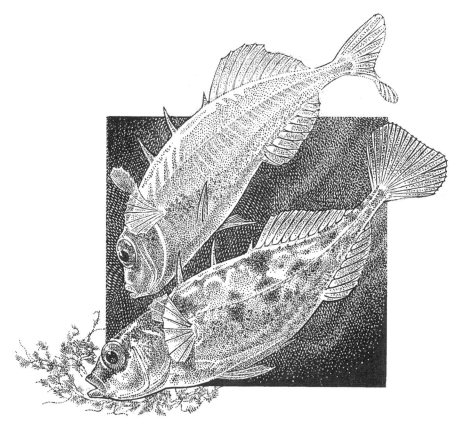

FIGURE 7.6   The three-spined stickleback, *Gasterosteus aculeatus*, a nest-tending evolutionary cousin of pipefishes and seahorses.

the evolutionary stage for females to outpace males in reproductive capacity. Over evolutionary time, as the eggs and embryos became housed securely inside sires, these gender biases were exaggerated, eventuating in the situation observed today, in which female pipefish often experience sexual selection as they compete for mating access to the limiting reproductive resource: available brood space inside receptive males.

Factoid: Did you know? A stickleback's kidney secretes a gluelike protein that the male uses to build his nest from vegetable material.

Brood Pouch Designs

The structure of the male's brood pouch varies greatly among extant syngnathid species. At one end of the spectrum are fortified brood pouches with physical complexity (often including placenta-like features). This condition is characteristic of extant seahorses. At the other end of the spectrum are relatively simple, unprotected brooding areas on a male's ventral surface, where eggs are glued but not encased. This condition is found in a few pipefish species. Between these two extremes are other pipefish species with either thin membranous compartments surrounding each egg or partially enclosed ventral pouches with protective coverings that extend across multiple eggs and embryos.

Wilson et al. (2003) address the evolutionary history of male brood pouches by mapping the phylogenetic distributions of alternative pouch designs on an mtDNA phylogeny for more than 30 syngnathid species (fig. 7.7). In this example of phylogenetic character mapping, a good agreement was found between clade membership and brood-pouch morphology. For example, all surveyed members of the *Syngnathus* pipefish clade possess closed pouches with two bilateral skin folds grown together, whereas seahorse species in the *Hippocampus* clade were unique in having a full saclike pouch enclosed by a single covering. On the other hand, some other pouch designs recurred in separate branches of the phylogenetic tree, thus suggesting independent evolutionary origins. Overall, the molecular phylogeny was consistent with the idea that pouches with simple designs evolutionarily predated more complex pouches.

One might suppose that male pregnancy itself should qualify syngnathids as being sex-role reversed relative to mammals. However, much of the scientific literature defines sex-role reversal as occurring whenever sexual selection operates more intensely on females than on males. By this criterion, some syngnathid species are sex-role reversed, but others are not. In some syngnathid species, females collectively produce many more eggs than males can accommodate in their brood pouches, thus making males a limiting reproductive resource for which females presumably compete. Sex-role reversal by this definition has other ramifications: females in such species should have polyandrous mating tendencies and should be the sex that displays sexually selected phenotypes. All of these properties differ diametrically from the biological situation in most other species, in which females are a limiting resource such that males experience stronger sexual selection and evolve elaborate phenotypes for attracting mates or battling among themselves for female access.

A vast scientific literature shows that the topics of sexual behavior, gender ratios, sexual selection, sexual dimorphism, and mating systems are all intertwined. To sort through some of this complexity for syngnathids, Wilson

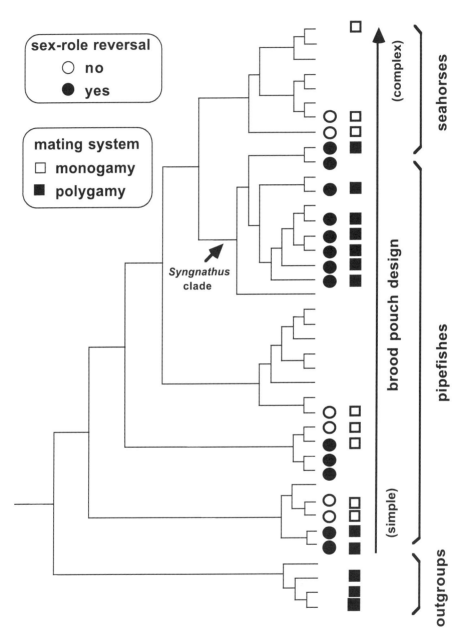

FIGURE 7.7 Molecular phylogenetic tree for 36 syngnathid species and four outgroup taxa (after Wilson et al. 2003). Mapped onto terminal nodes of the tree are the distributions of monogamous and polygamous mating systems and presence versus absence of sex-role reversal among extant species.

et al. (2003) also used their molecular phylogeny (see fig. 7.7) to address the evolutionary histories of sex-role reversal and alternative mating systems in male-pregnant fishes. One possibility the authors explored was that fancy brood pouches predict sex-role reversal because males with such high-investment pouches might be an even more limiting resource to mate-prospecting females. However, this possibility was not confirmed by the PCM analyses, which showed instead that sex-role-reversed clades were present in both simple-pouch and complex-pouch regions of the syngnathid phylogeny. Perhaps different brood-pouch designs are not reliable indicators of male investment in offspring care.

The PCM analyses by Wilson et al. (2003) do, however, support another evolutionary hypothesis: that sex-role reversal in syngnathid species tends to be associated with polyandry. For example, all *Syngnathus* pipefishes for which genetic information is available are both polygamous and sex-role reversed (see fig. 7.7), whereas all such *Hippocampus* seahorse species are monogamous and lack sex-role reversal according to the sexual-selection definition.

Genetic-Parentage Analyses

Male pregnancy, a phenomenon foreign to mammals, affords novel vantage points on animal reproductive behaviors. Apart from PCM, another comparative approach applied to numerous syngnathid species (Jones and Avise 2001) has involved genetic-parentage analyses via molecular markers (see the appendix). The general intent of this approach is to address the relation of sexual selection and sexual dimorphism to alternative genetic mating systems. For example, in some or all species of male-pregnant syngnathids, polyandry might be common, and sexual selection on females (via competition for mates) might actually be more intense than on males such that females evolve brighter colors or other special body ornaments. Indeed, during the breeding season females in many pipefish species become festooned with secondary sexual traits, a reversal of the typical situation in birds and mammals, in which males are the adorned sex. In theory, the keenness of female competition for mates should also relate to the genetic mating system of each pipefish and seahorse species. For example, sexual selection on females might be most intense in polyandrous species because any female who successfully woos several choosy males might markedly improve her genetic fitness (especially if she can produce more eggs than a male can typically incubate).

Almost all of the genetic-parentage analyses on syngnathids were conducted on broods of pregnant males collected from nature. Examples of the kind of maternity data that emerge from such analyses are shown in figure 7.8. As expected (given intrapouch fertilization), no instances of cuckoldry were detected in the dozens of syngnathid litters analyzed, meaning that

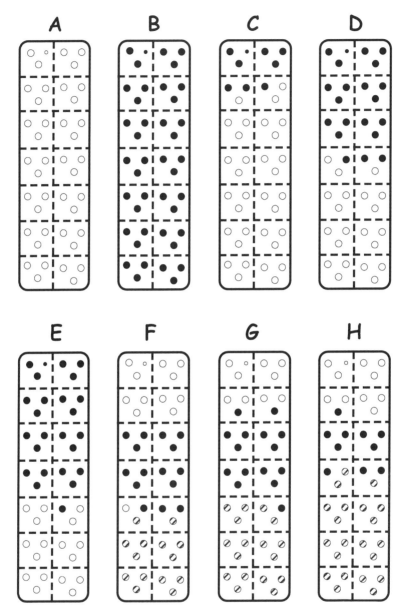

FIGURE 7.8 Diagrams of results from genetic-parentage analyses as applied to pipefishes (after Jones and Avise 1997b). Shown are embryos (small circles) within the brood pouches (large rectangles) of eight pregnant *S. floridae* males. Each pregnant male proved to have sired all of the embryos within his pouch, and 1–3 females contributed to a given brood. For example, the brood pouch of sire *H* housed embryos from three dams (open, closed, and lined circles), whereas sires *A* and *B* had only one successful mate each. Such data accumulated for many broods helped to reveal the population's genetic mating system.

Factoid: Did you know? Whenever a pregnant pipefish male proved (by genetic-parentage analyses) to have mated with multiple females, the different full-sib clutches in his pouch were arranged in a neatly stacked configuration, suggesting that eggs received from successive mating events sink toward the bottom of his pouch.

each pregnant sire has a very high assurance of paternity for the offspring he carries. This certainty of paternity in the Syngnathidae in turn facilitated molecular analyses of biological maternity and genetic mating systems (because the dam's genotype for each gestating embryo could be determined by subtracting known paternal alleles from each offspring's diploid genotype).

From this type of molecular information on biological parentage, several different genetic mating systems ranging from monogamy to polygynandry to polyandry were uncovered in different syngnathid species (Jones and Avise 2001). Furthermore, when these genetic mating systems were interpreted in the context of observed levels of sexual dimorphism and the presumed intensities and directions of sexual selection in the surveyed species, the overall results appeared compatible with conventional wisdom (as summarized in fig. 7.1) for taxa that include species with proclivities toward polyandry and sex-role reversal. Thus, those assayed syngnathid species that proved to have polyandrous or polygynandrous genetic mating systems displayed greater sexual dimorphism than did the monogamous species. Furthermore, as might have been predicted for the sexually dimorphic syngnathids, it was invariably the female gender that exhibited the more pronounced development of secondary sexual characteristics.

Another interesting finding emerged from genetic analyses of seahorses. From behavioral observations in nature, most seahorse couples seem highly devoted to one another. However, genetic markers proved that the progeny in a male's successive broods often had different mothers. Evidently, seahorse couples sometimes "divorce" and "remarry" a variety of partners (yielding a mating system known as *serial monogamy*). In other words, although each pregnancy consisted of full-sib offspring (evidencing monogamy), individuals frequently switch partners during or between breeding seasons (in effect evidencing polygamy over time).

Factoid: Did you know? In some seahorse species, a mated male and female greet each other daily with ritualized behaviors that include head nods plus hugs with their prehensile tails.

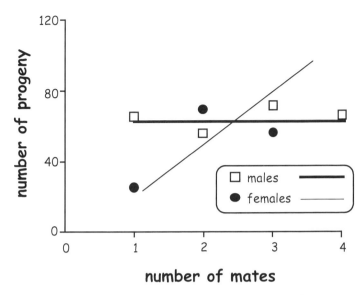

FIGURE 7.9   Sexual selection gradients empirically documented in a male-pregnant pipefish species, *Syngnathus typhle*. Note that slopes in the regressions for male and female reproductive success are reversed (steeper for females) compared to those that typify female-pregnant and many other species (see box 7.1).

Finally, some of the genetic-parentage analyses in the pipefish *Syngnathus typhle* were conducted under controlled laboratory conditions where sex ratios and mating opportunities were manipulated experimentally (Jones et al. 2000, 2005). The most salient finding was that females in this polygynandrous pipefish species show higher fitness gains from multiple mating than do males. Thus, Bateman gradients in this male-pregnant species indeed are reversed (fig. 7.9) compared to the standard situation in female-pregnant species and other taxa with conventional sex roles.

External Pregnancy

Fish display diverse reproductive modes ranging from pelagic group spawning to social monogamy and from internal female pregnancy (as in Poeciliidae) to internal male pregnancy (in Syngnathidae). Collectively, this diversity provides rich fodder for comparative evolutionary investigations of alternative reproductive strategies (Avise et al. 2002). Of special relevance to this book on pregnancy is the topic of parental care, which across fish species ranges from nonexistent to extensive. Approximately 89 of 422 taxonomic families of bony fish (21%) contain at least some species that practice parental

FIGURE 7.10   The nurseryfish, *Kurtus gulliveti*, an Australian species in which each male uses a special hook on his forehead to carry clusters of eggs (Berra and Humphrey 2002). How the eggs get transferred from the female (and fertilized) remains unknown.

care of offspring, and, interestingly, in nearly 70% of such cases the primary or exclusive parental custodian is the male (Blumer 1979, 1982). Parental care in various fish species includes not only commonplace phenomena like nesting and oral brooding but also some rare and bizarre mechanisms such as that displayed by the nurseryfish (fig. 7.10). Exclusive paternal care of offspring is otherwise quite uncommon in the biological world, so fish offer mirror-image evolutionary perspectives on parental care compared to many other animal groups where females are the primary caregivers (Clutton-Brock 1991).

Oral Brooding

Mouth brooding (or sometimes gill brooding) occurs in several piscine groups, including catfishes, lumpfish, cardinal fish, jawfish, and some blind cave fishes. However, nowhere has the phenomenon been studied more thoroughly than in the Cichlidae (fig. 7.11).

Factoid: Did you know? Some predatory cichlid fish ram the head region of a mouthbrooder, forcing the latter to spit out a few offspring, which the predator then eats (McKaye and Kocher 1983).

FIGURE 7.11 Mouth brooding behavior in cichlid fishes. Adults in many species in the family Cichlidae routinely safeguard eggs and young in their mouths. Depending on the cichlid species, oral brooding may be by sires, dams, or both parents.

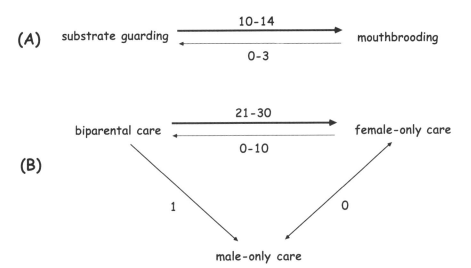

FIGURE 7.12 Evolutionary transitions (as deduced from PCM analyses) between alternative modes of parental care of offspring in cichlid fishes with external fertilization (after Goodwin et al. 1998). Shown are the numbers of independent transitions between the following: (A) substrate guarding and mouth brooding; and (B) biparental, female-only, and male-only care.

Phylogenetic character mapping has focused on evolutionary transitions among alternative parental-care modes (including mouth brooding) in species representing nearly 200 cichlid genera (Goodwin et al. 1998). The results demonstrated that female-only care evolved many times from biparental care, whereas male-only care rarely did so (fig. 7.12). Results further showed that substrate guarding was the ancestral mode of parental care in Cichlidae, from which mouth brooding evolved on at least 10 occasions (fig. 7.12). Perhaps the transport of offspring between excavated pits by a substrate guarder provided a first step toward the evolution of mouth brooding, especially if the offspring were also retained in the adult's mouth for some time to avoid predation (Baylis 1981). Additional research has shown that evolution of mouth brooding in cichlids is also associated with the loss of adhesive threads on eggs, reduced fecundity, increased egg size, and higher juvenile survival (Keenleyside 1991).

Nest Tending by Males

As described in chapter 2, when alternative modes of parental care are mapped onto molecular phylogenies of fishes, it becomes apparent that paternal care has arisen many times during piscine evolution, typically from ancestral states of external fertilization without parental care (see fig. 2.3). The evolution of any such gender-specific parental care in species with external fertilization is usually explained by the "proximity argument," wherein the gender that is closest to the eggs at the time of fertilization tends to be the sex that is predisposed to care for the young, especially if the other sex then deserts or abandons the effort for any reason (Trivers 1972; Williams 1975; Dawkins and Carlisle 1976). Especially in fish species that practice male territoriality and nest guarding, the gender that is most proximate to the zygotes and the early embryos is typically the male. This proximity factor undoubtedly helps to account for the prevalence of male-only care in nest-tending fish species.

Especially when the magnitude and duration of paternal care are substantial, many nest-tending fish species display what can be thought of as "external male pregnancy" (Avise and Liu 2010, 2011). In such species, a "bourgeois" male typically constructs or adopts a nest, into which one or more females deposit eggs that the "shopkeeper" male may fertilize. Each bourgeois male then guards and tends the embryos for several weeks until the young leave the nest to begin independent lives. In effect, each nest-tending male assumes parental duties that in some ways parallel the time and energy expended by the gestating parent in viviparous fishes. Furthermore, genetic-parentage analyses (see appendix) can be applied to the broods of nest-tending fishes just as they can for broods in viviparous fish species with either internal male pregnancy or internal female pregnancy (fig. 7.13).

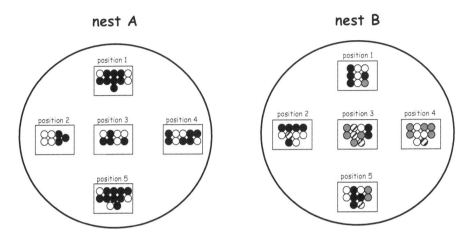

FIGURE 7.13   Examples of maternity revealed by molecular markers in a nest-tending fish species, *Lepomis auritus* (after DeWoody et al. 1998). Shown are embryos (small circles) at five sample locations within each of two nests tended by bourgeois males. The coded circles indicate that two females contributed to nest *A* and four females contributed to nest *B*. Such maternity assignments accumulated for many nests help reveal a population's genetic mating system, which also might include instances of cuckoldry.

Factoid: Did you know? Fish nests have varied designs ranging from simple saucer-shaped depressions in the substrate (as in the sunfishes) to elaborate tunnel nests built above the substrate and consisting of woven and glued plant materials (as in refined nesters such as sticklebacks). Gouramis and Siamese fighting fish make froth nests consisting of masses of mucus-covered bubbles. Another odd type of fish nest is described in the caption to figure 7.14. A few fish species even lay their eggs in the nests of other nest-tending fish species (Helfman et al. 1997).

There are, nevertheless, several key differences between internal pregnancy and external pregnancy via nesting. For example, external pregnancy presumably entails fewer constraints on brood space (and therefore on progeny numbers) than does internal pregnancy, and this might have major consequences for the operation of sexual selection (chapter 8) and the evolution of piscine mating phenotypes and behaviors. External pregnancy also opens windows of opportunity for "stolen fertilizations" (cuckoldry events) that otherwise remain closed to viviparous species with internal fertilization. Such exposure to cuckoldry in nest-tending fishes in turn produces selection

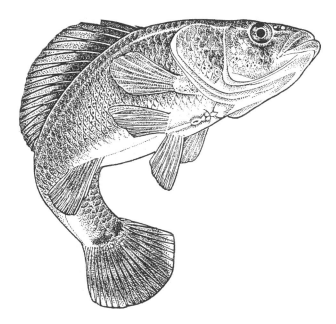

FIGURE 7.14   The pirate perch, *Aphredoderus sayanus*. In this freshwater species from swamps in the eastern United States, adults lay and fertilize eggs in tunnels excavated (by invertebrates) in the submerged balls of tree roots. The peculiar anterior position of the urogenital pore in the pirate perch allows each fish to enter a tunnel headfirst and spray its gametes into the root ball's protective interior, where the zygotes and embryos then incubate without further parental involvement (Fletcher et al. 2004).

pressures that have led to an amazing array of "alternative reproductive tactics" (ARTs) by males (Taborsky 1994; Gross 1996; Henson and Warner 1997; Thomaz et al. 1997; Oliveira et al. 2008).

Cuckoldry and Alternative Reproductive Tactics

To introduce the topic of cuckoldry in fishes, consider the bluegill sunfish (fig. 7.15), a nest-tending species with several ARTs. Three types of bluegill males were found in surveyed populations of this species in Canada: (a) bourgeois males, which mature at about 7 years of age and then construct saucer-depression nests in colonies, attract and spawn with females, and vigorously defend the nest and embryos; (b) precocious sneaker males, 2–3 years of age, which dart into a bourgeois male's nest and release sperm; and (c) older satellite males, which mimic females in color and behavior but also release sperm as the primary couple spawns in a bourgeois male's nest (Gross and Charnov 1980). Genetic-parentage analyses (Colbourne et al. 1996; Neff 2001;

FIGURE 7.15   The bluegill sunfish, *Lepomis macrochirus*, one of many nest-tending species in the family Centrarchidae. In this species, various offspring in a nest might have been sired by the bourgeois male (the nest tender), sneaker males, or female-impersonator males (see text).

Philipp and Gross 1994) have shown that about 20% of the offspring in a bluegill colony are foster progeny resulting from cuckoldry (fig. 7.16). Studies also have suggested that bourgeois males can detect lost paternity and adaptively lower their level of parental care accordingly (Neff and Gross 2001). Cuckoldry has been genetically documented in several other sunfish species as well (fig. 7.16), albeit typically at lower rates than in bluegills.

"Extrapair fertilization" (EPF) events that underlie foster parentage in nest-tending species (including birds and fishes) are generally thought to increase the intensity of sexual selection because they presumably often increase the overall variation in favorable male mating outcomes, especially if (as is often true) successful bourgeois males are among the effective cuckolders (Møller and Birkhead 1994). However, Jones et al. (2001) uncovered a

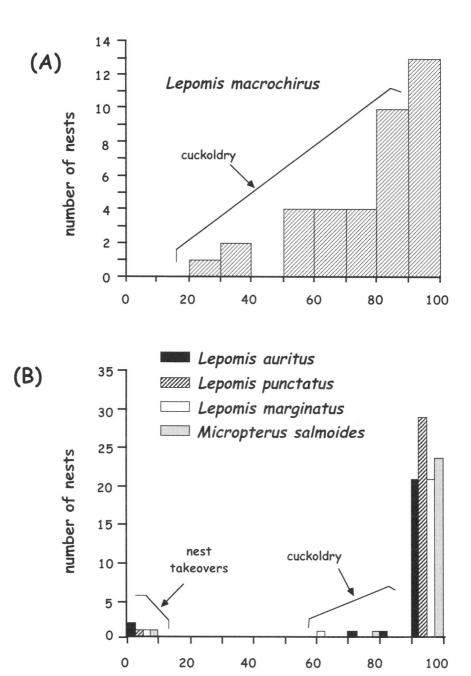

FIGURE 7.16 Molecular findings on genetic paternity in several species of sunfish (after Avise et al. 2002). Shown are percentages of progeny per nest that proved to have been sired by bourgeois males in: (A) 38 nests of *Lepomis macrochirus*; and (B) 104 nests of four other centrarchid species.

converse case in which EPFs and ARTs probably decreased the intensity of sexual selection. For a marine goby (*Pomatoschistus minutus*), the authors showed by genetic-parentage analyses that sneaker males had fertilized eggs in nearly 50% of the assayed nests but also that, as a result, the overall inter-male variance in reproductive success was lower than it probably would have been in the absence of such high rates of cuckoldry.

Similar molecular appraisals of genetic-parentage and mating systems in a wide variety of nest-tending fish species (see appendix) have been used in conjunction with behavioral observations to reveal other oft-cryptic repro-ductive phenomena, including those described in the following sections.

Multiple Mating by Bourgeois Males

A given fish nest has often proved to contain half-sib offspring from multiple (typically 2–8) dams (DeWoody and Avise 2001). Such genetic evidence of multiple mating by bourgeois males has been gathered for several species of *Lepomis* sunfish (DeWoody, Fletcher, et al. 1998, 2000b; Mackiewicz et al. 2002), *Etheostoma* darters (DeWoody, Walker, et al. 2000; Porter et al. 2002), *Spinachia* sticklebacks (Jones, Östlund-Nilsson, et al. 1998), *Pomatoschistus* sand gobies (Jones et al. 2001), and *Cottus* sculpins (Fiumera et al. 2002), among others.

Other Routes to Foster Parentage

Two other behavioral avenues to nonpaternity by bourgeois males involve occasional nest takeovers and instances of egg piracy, which are provision-ally evidenced when few or no offspring in a focal nest prove to have been genetically sired by the resident bourgeois male. Nest takeovers might be opportunistic responses by males to limited nest-site availability, or perhaps a nest holder captured at the time of sampling was merely a temporary visi-tor at the nest (e.g., there to cannibalize the embryos of another male). Egg thievery or nest raiding (i.e., the theft of a few eggs from a neighbor's nest) has been documented in sticklebacks (Jones, Östlund-Nilsson, et al. 1998; Li and Owings 1978; Rico et al. 1992). Such egg raiding might seem counterin-tuitive from an evolutionary perspective, but one plausible explanation is that the behavior benefits the thief by seeding or priming his own nest with eggs that are known to be effective in many fish species in eliciting spawning responses by additional females, with whom the resident male then mates (review in Porter et al. 2002).

Egg Mimicry

For males of many fish species, various body parts or pigmented regions have evolved a close resemblance to conspecific fish eggs. In fantail darters

FIGURE 7.17   One of several darter species (family Percidae) in which body parts of the male have evolved to resemble fish eggs. In this case, egglike structures are located on spine tips of the male's anterior dorsal fin.

(fig. 7.17), each spine on the foremost dorsal fin carries one round glob that looks just like a fish egg. In other darter species, these egglike globs are located on the spiny tips of a posterior dorsal fin, and in yet another group of darters they reside on pelvic fins near the male's throat. Thus, egg mimicry structures have evolved independently several times in various darter lineages (Page and Bart 1989). In many darter species, males nest in cavelike openings in a streambed, where females cement eggs to the cave's ceiling or walls. Primitive versions of these fleshy protuberances in ancestral darters may have served to blunt a bourgeois male's sharp spines, which otherwise could puncture some of the eggs that line his nesting chamber (Page and Swofford 1984). Only later in evolution did the structures presumably become refined and useful also as egg mimicry devices that might help a male to attract females to lay bona-fide eggs in his nest.

For many nest-tending and mouth-brooding fish species, it is well documented that females prefer to spawn with males that seem to display or tend eggs (Knapp and Sargent 1989; Porter et al. 2002). This might be because such eggs are an indicator of good health in a potential mate, or they may speak favorably of a male's parenting skills. Indeed, this is one reason that occasional egg thievery and nest piracy in nest-tending fishes (as described earlier) might make evolutionary sense after all. By stealing a few fertilized eggs or pirating a nest, a male may be priming a reproductive pump that brings additional unfertilized eggs into his nest from gravid females who are impressed with his apparent parenting potential.

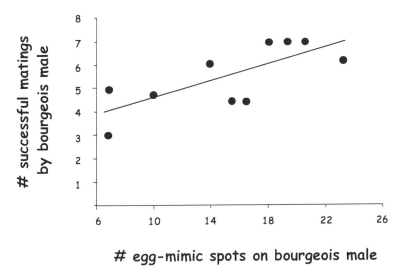

egg-mimic spots on bourgeois male

FIGURE 7.18   Observed correlation ($r=0.74$, $P<0.05$) between numbers of egg-mimic spots and genetically deduced numbers of mates secured by nest-tending striped darter males (after Porter et al. 2002).

Factoid: Did you know? Egg-mimic spots can have other functions, too. In some mouth-brooding cichlids, males on their anal fins have "egg dummies" to which females are attracted. As the female mouths the dummies (just as she does her own bona-fide eggs), the male releases sperm and fertilizes the eggs that she carries within her mouth.

Factoid: Did you know? "Sperm drinking" is an even more bizarre form of insemination displayed by armored catfishes in the family Callichthyidae. In these species, a female swallows a male's sperm, quickly passes the ejaculate through her digestive system, and then releases still-viable sperm to fertilize eggs that she holds between her pelvic fins (Kohda et al. 1995).

In most darter groups, the number of egg-mimic structures per male is essentially constant because each fish has a fixed number of fin rays. However, in one population of the striped darter (*Etheostoma virgatum*), the number of egg look-alikes varies from half a dozen to more than 20 per male. This is because, unlike the physical egg-mimic structures in the other darter species, these egg look-alikes are unpigmented spots on the darker background of the male's pectoral fins. Because of this variable number of faux eggs on bourgeois males, *Etheostoma virgatum* offered researchers an opportunity to test the egg-mimicry hypothesis empirically. Do males with more mimetic spots spawn with more females than do males with fewer spots? To answer this question, Porter et al. (2002) used genetic maternity assignments for dozens of embryos in the nests of several striped darter males with varying numbers of egglike spots. Their results supported the notion that females spawn preferentially with males that display more of these spots (fig. 7.18).

Perhaps this is an honest display—male darters with many egg look-alikes might be healthier or of higher genetic quality on average and are merely advertising that truthful fact to females. On the other hand, the egg-mimic spots might be deceitful displays by males, signifying little or nothing about their true parenting abilities. This demonstrates that deceptive and honest displays can both be the objects of various forms of sexual selection in fishes and other species.

Filial Cannibalism

Cannibalism (eating other members of one's own species) is a common phenomenon in the animal world and often makes good sense because successful cannibals may thereby increase their own survival and reproduction at the expense of their competitors for food, space, or mates. But why might a parent eat its own offspring as well, given that progeny carry copies of its parent's genes? Yet "filial cannibalism" seems to occur in nest-tending fish species, where a guardian male is occasionally observed eating a few eggs or fry from the nest he is tending (Fitzgerald 1992). Filial cannibalism is different from "heterocannibalism" (the consumption of conspecific nonrelatives). Why might a sire consume some of his own offspring?

Several explanations are possible. Perhaps natural selection favors filial cannibalism when a nest-guarding male is otherwise facing starvation. Although clearly detrimental to those who are eaten, such cannibalism may be beneficial to the male and to his remaining kin if at least some members of the current and subsequent broods survive. A second possibility is that the eggs or fry that a male eats from his nest are infected with a fungus or are otherwise diseased and would not survive anyway. Indeed, by eating infected eggs, a sire might stop the spread of disease within his nest. A third possibility is that filial cannibalism is simply a nonadaptive behavioral byproduct of a fish's voraciousness. In terms of the final tally of offspring produced, it probably matters little if a few babies are consumed from a nest that may contain scores or hundreds of eggs and fry.

A fourth viable hypothesis (especially given the high incidence of cuckoldry, as described early) is that a nest-tending male is actually ingesting foster progeny rather than his own. Because many fry in a nest now are known

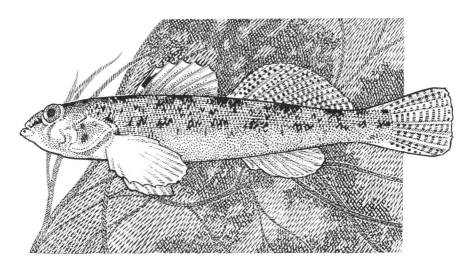

FIGURE 7.19   The tessellated darter, *Etheostoma olmstedi*, a species in which males tend nests in gravel streambeds and sometimes eat a few of their own offspring.

from genetic-paternity analyses to be tended by a foster father, might some purported instances of filial cannibalism actually register as heterocannibalism instead? To address this possibility, DeWoody, Fletcher, et al. (2000) genetically examined freshly eaten embryos retrieved from the stomachs of cannibalistic male tessellated darters (fig. 7.19) who were tending their nests. From genetic paternity analyses conducted on dozens of partially digested embryos, these researchers proved that the bourgeois males indeed had eaten some of their own biological progeny. Thus, filial cannibalism is a real phenomenon after all.

SUMMARY

1. In contradistinction to natural selection that forges adaptations to environmental demands, sexual selection molds behaviors and other phenotypes to mating demands. Woven into a tapestry of topics, sexual selection includes mating behaviors, mating systems, sex ratios, sexual dimorphism, and other sexual phenomena that ultimately trace their evolutionary geneses and gender biases to anisogamy. Female pregnancy also fits into this conceptual framework since it is both a logical outgrowth of anisogamy and an amplifier of anisogamy's effects, especially with regard to sexual selection.

2. Sexual selection can be dichotomized in several ways: intrasexual/intersexual; precopulatory (typically via female mating preferences)/postcopulatory

(via sperm competition and/or cryptic female choice); association with alternative mating systems, including monogamy and polygamy (polygyny, polyandry, and polygynandry); or operating primarily on males rather than females. Such characterizations are not mutually exclusive, but they do highlight the many overlapping ways in which sexual selection can operate.

3. Fish are excellent subjects for comparative appraisals of sexual selection because they include many representative species with internal female pregnancy, internal male pregnancy, and external male pregnancy (e.g., nest tending). The evolutionary origins and selective ramifications of different forms of pregnancy have been addressed using three broad approaches: phylogenetic character mapping (PCM) to deduce the evolutionary histories of particular pregnancy-related traits; field observations and experiments to unveil the functional significance of those traits; and molecular-parentage analyses to reveal the genetic (as opposed to the "social") mating systems of particular fish species. The strongest conclusions typically emerge from successful mergers of several of these research approaches.

4. Molecular-parentage analyses are especially germane because each pregnant adult is physically associated with its brood, and this association greatly facilitates deductions about who spawned the brooded progeny. Genetic-parentage analyses have been conducted on many broods in a wide variety of viviparous fishes in which either females (e.g., in Poeciliidae) or males (in Syngnathidae) become pregnant. Genetic analyses have been likewise conducted in numerous nest-tending and other fishes with external fertilization and external male pregnancy. A complication of these latter species is that a bourgeois (nest-tending) male has sometimes been cuckolded via extrapair fertilizations. Genetic markers have confirmed that foster parentage also arises occasionally in nest-tending species via egg thievery and nest piracy. Additional phenomena addressed by genetic-parentage analyses include egg mimicry and female choice, filial cannibalism, alternative reproductive tactics (ARTs), Bateman gradients, and sexual selection in relation to alternative expressions of piscine pregnancy, sexual dimorphism, and genetic mating systems.

CHAPTER EIGHT

Pregnancy in a Comparative Light

We learned in chapters 1 and 7 that anisogamy (the larger size and lower motility of female gametes) initiated an evolutionary cascade of gender biases
with respect to potential fertilities, variances in reproductive success, intensities of mate competition, the nature and direction of sexual selection, the
elaboration of secondary sexual traits, magnitudes of parental investment in
progeny, proclivities for pregnancy and brooding, and assurances of biological parentage for particular offspring. We also learned that female pregnancy
in viviparous taxa then often amplifies these sexual biases by further curbing
female fecundity and making each female even more limiting as a reproductive resource. Female pregnancy then feeds back into the entire procreative
processes by synergistically boosting the sexual biases underlying sexual selection and mating strategies. Female pregnancy also affects the way in which
natural selection and kin selection jointly influence the evolutionary trajectories of phenomena such as genomic imprinting, immunological reactions,
and genetic conflicts between parents, among siblings, and between offspring
and parents (chapter 6). Figure 8.1 gives a diagrammatic overview of this complex nexus of evolutionary causation.

Having addressed many of the selective forces at work during mammalian (chapter 6) and piscine (chapter 7) pregnancies, we can now compare
gestational phenomena across these and other brooding taxa from a broader
evolutionary vantage point. This final chapter examines the impact of alternative gestational modes (such as male pregnancy/female pregnancy and
internal/external brooding) on sexual selection and natural selection.

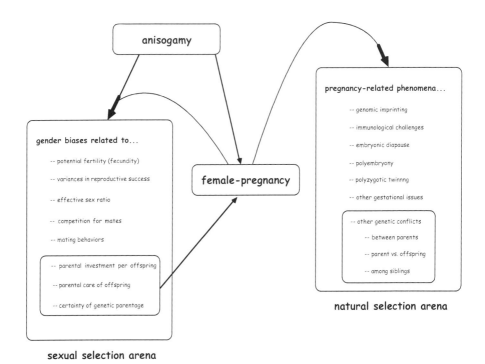

FIGURE 8.1   The central places of anisogamy and pregnancy in the broader evolutionary framework of sexual selection and natural selection related to embryonic gestation. Arrows show just some of many causal connections and feedback loops (see text).

Sexual Selection and Multiple Mating by the Pregnant Gender

As Bateman (1948) famously emphasized (chapter 7; see also Wade and Shuster 2010), a male can in principle increase his reproductive output dramatically by mating with several or many females, whereas a female normally stands to gain much less from taking multiple mates. Consequently, males in many outcrossing species (viviparous or otherwise) typically experience stronger intra- and intersexual selection than do females and thereby tend to evolve and display greater elaborations of secondary sexual traits.

Further support for Bateman's insights into causal relationships among mate numbers, reproductive fitness, and sexual selection has come from appraisals of genetic mating systems in pipefishes and seahorses (Syngnathidae), which display the peculiar phenomenon of internal male pregnancy. As detailed in chapter 7, some pipefish species show sex-role reversal in the sense that Bateman gradients are steeper for females than for males, polyandry is common, sexual selection operates more intensely on females, and females tend to evolve secondary sexual characteristics. Additional contrasts of this sort are made possible by the many fish species in which bourgeois males tend embryos in nests. With respect to parental investment in offspring, such species in effect display external male pregnancy (chapter 7). Similarly, in a few invertebrate taxa, including Pycnogonida (sea spiders) and Belostomatidae (giant water bugs), males alone brood the young, in this case on their outer body surfaces (chapter 4). Such observations raise this question: How do alternative forms of pregnancy affect the evolution of mating behaviors by the parental sex that conducts the brooding (as well as by the opposite gender)?

As applied to viviparous and other species with extended parental care of progeny, Bateman's (1948) reasoning implies that any member of the non-pregnant sex can normally enhance its reproductive output by mating with multiple partners, whereas much more problematic is the degree to which a member of the pregnant sex might profit from multiple mating. Thus, at issue in female-pregnant species is not why males routinely seek multiple mates, but how often and why females might also do so.

In theory, multiple mating by females might confer any of several fitness benefits on a polyandrous dam in female-pregnant taxa. Hypotheses to account for polyandry generally fall into two broad categories (table 8.1) depending on whether the purported benefits to a female are direct (material) or indirect (genetic). Examples of direct benefits include the receipt of courtship gifts, better access to male-held territories, a reduced risk of sexual harassment from males, and improved chances of finding a behaviorally compatible mate. Examples of indirect benefits to a polyandrous female include a better opportunity to receive "good" paternal genes for her offspring, higher genetic diversity within her brood, fertilization insurance against the possibility that some males are sterile, hereditary "bet hedging" in unpredictable environments, and/or several other potential bonuses that might elevate a polyandrous female's fitness above the fecundity plateaus otherwise imposed by anisogamy and pregnancy. Yet another possibility is that polyandry is merely a correlated evolutionary response in females to positive selection for the same genes that underlie polygyny in conspecific males (Forstmeier et al. 2011).

For several reasons the topic of multiple mating by members of the pregnant sex has attracted scientific interest in recent years. First, viviparity is a high-investment tactic that seems likely to promote strong selective pressures on mating proclivities by adult brooders. Second, molecular markers have been deployed to illuminate the genetic mating systems of many viviparous and brooding species in nature (see the appendix). Third, these genetic parentage analyses are ideally suited for estimating successful mate numbers for pregnant individuals because each litter comes conveniently "prepackaged"

Fishes with Alternative Pregnancy Modes

Empirical findings from extensive genetic-parentage analyses of piscine broods generally support many of the predictions listed earlier, albeit sometimes only weakly. In a survey of the scientific literature on genetic mating systems in fish species with female pregnancy, internal male pregnancy, or external male pregnancy (Avise and Liu 2010), successful polygamy by members of the brooding sex proved to be common (fig. 8.4), as more than 50% of all broods comprised two or more (and as many as nine) full-sib cohorts. This is perhaps unsurprising, given the many potential fitness payoffs of polygamous matings (especially to members of the nonpregnant sex). Furthermore, rates of multiple mating and mean mate counts were significantly higher in species with external pregnancy than they were for species with internal pregnancy, and they were also higher for dams in internal female-pregnant species than they were for sires in internal male-pregnant species (fig. 8.5). These empirical outcomes are all consistent with the notion that different gestational modes have rather predictable consequences for each parent's effective fecundity and thereby for its exposure to selection for polygamy. On the other hand, all of these trends were statistically mild at best. Indeed, perhaps the more telling revelation of this empirical overview of piscine genetic mating systems was the surprisingly small mean number (typically only 1–4) of full-sib cohorts per brood across a wide diversity of fish species with very different pregnancy modes.

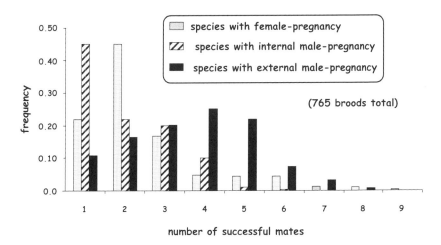

FIGURE 8.4   Genetically deduced numbers of successful mates that contributed to each of 765 broods of 29 pregnant fish species (after Avise and Liu 2010).

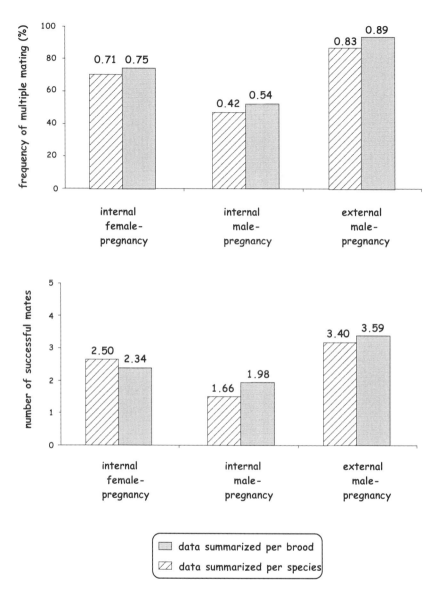

FIGURE 8.5 Genetically deduced frequencies of multiple mating (above) and mean numbers of successful mates per brood (below) for parents that were incubating a total of 765 broods in 29 fish species, representing three alternative types of pregnancy (after Avise and Liu 2010). Shown are mean values as calculated both on a per-species (regardless of brood number) and per-brood basis.

Fishes Versus Mammals

To further address the evolutionary ramifications of finite brood size and truncated fecundity for mating behaviors in the pregnant sex, Avise and Liu (2011) expanded this review of the genetic-parentage literature to include

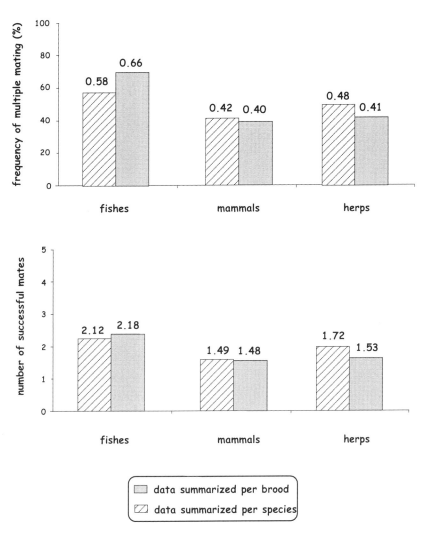

FIGURE 8.6   Genetically deduced frequencies of multiple mating (above) and mean numbers of mates per brood (below) for parents that were incubating a surveyed total of 533 broods of viviparous fishes; 1,930 mammal broods; and 362 broods of viviparous herps (after Avise and Liu 2011).

viviparous amniotes (mammals and reptiles) and live-bearing amphibians. As was also true for the pregnant fishes, multiple mating by the pregnant sex proved to be common in most other live-bearing vertebrates (fig. 8.6). Furthermore, as might have been predicted given their 10-fold smaller clutch sizes, these other vertebrate groups tended to show lower rates of multiple mating and averaged fewer mates per brood than did their pregnant piscine counterparts. However, these trends again were modest and gave little evidence that clutch size per se is a key selective factor underlying mating decisions by individuals of the pregnant sex. Instead, the deduced numbers of full-sib cohorts per brood typically evidenced only about 1–4 genetically successful mates per brood, regardless of brood size. Furthermore, only weak correlations were detected between brood size and mate numbers within (and among) these three vertebrate groups (fig. 8.7).

Invertebrates Versus Vertebrates

To extend such analyses further, Avise et al. (2011) then expanded their review of the genetic-parentage literature (see appendix) to include invertebrate animals that brood their young. As was also true for pregnant vertebrates, multiple mating proved to be common in the surveyed invertebrate brooders (fig. 8.8). Indeed, brooding invertebrates showed higher frequencies of polygamy and averaged more mates per brood than did their pregnant vertebrate counterparts, as might have been predicted given their much larger clutch sizes (often numbering in the hundreds to thousands of brooded embryos). Nevertheless, these tendencies again were modest and gave little evidence that clutch size per se has been a key selective factor governing mating decisions by members of the brooding sex. Instead, genetically documented numbers of mates per brood typically ranged upward to only about half a dozen individuals even in invertebrate taxa with extraordinarily large broods. This latter finding is intriguing because such huge broods could in principle house dozens or even hundreds of different full-sibships and also because molecular markers employed in the genetic surveys typically were polymorphic enough to have documented many more parents, had they in fact contributed to a given brood.

These findings from genetic-parentage analyses indicate that factors other than brood-space constraints must truncate multiple mating in nature far below levels that could otherwise be accommodated in most brooding invertebrates. Thus, these findings generally parallel the conclusions reached earlier for the pregnant vertebrates.

Overall, these comparative genetic appraisals of mating proclivities by individuals that become pregnant or otherwise brood their young have highlighted an interesting research irony. Whereas a stated goal in many theoretical and empirical studies of animal mating systems (see reviews

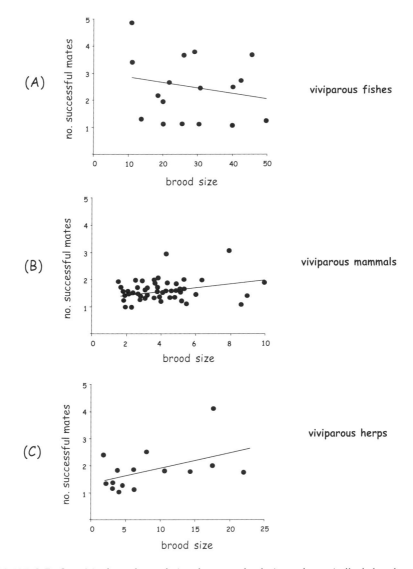

FIGURE 8.7 Surprisingly weak correlations between clutch size and genetically deduced mean mate numbers per viviparous brood for: (A) 17 fish species ($r=0.31$; $P=0.23$); (B) 49 mammal species ($r=0.27$; $P=0.05$); and (C) 15 species of reptiles and amphibians ($r=0.52$; $P=0.04$) (after Avise and Liu 2011).

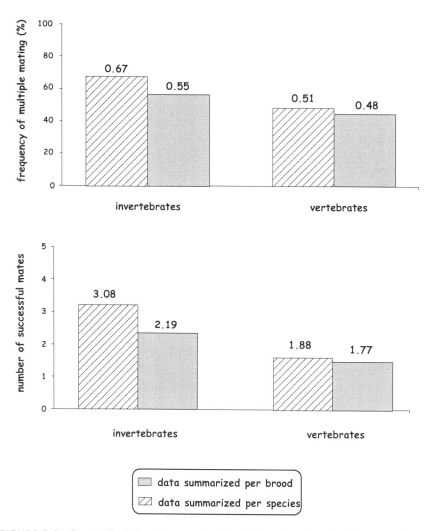

FIGURE 8.8   Genetically deduced frequencies of multiple mating (above) and mean numbers of successful mates per brood (below) for a total of 3,057 broods in 93 pregnant vertebrate species versus 583 broods representing 29 invertebrate species (after Avise et al. 2011).

in Arnqvist and Nilsson 2000; Simmons 2005) has been to understand why a gestating parent (typically the female) has so many mates and such a high proclivity for polygamy (see table 8.1), an oft-overlooked but equally compelling question is exactly the converse: Why do brooders and pregnant parents typically have so few successful mates?

Part of the answer probably has to do with costs to females of "too much" mating. For example, in their discussion of mating behaviors in promiscuous marine snails, Johannesson et al. (2010) noted that whereas "male fitness is expected to increase with repeated matings in an open-ended fashion, female fitness should level out at some optimal number of copulations." However, apart from the diminishing returns and costs of multiple mating, much of the explanation for the observation that members of the pregnant sex generally have so few sexual partners probably has to do simply with the myriad logistical constraints on mating. In nearly every species, ecological and behavioral hindrances to successful mating abound (Hubbell and Johnson 1987; Gowaty and Hubbell 2009; Avise and Liu 2011). Depending on the population, restraints on mate acquisition might include any of numerous factors such as low population densities, short mating seasons, low mate-encounter rates, lengthy courtships, and perhaps even the postcopulatory phenomena of sperm competition and cryptic female choice (Jones and Ratterman 2009; Eberhard 2009; Birkhead 2010).

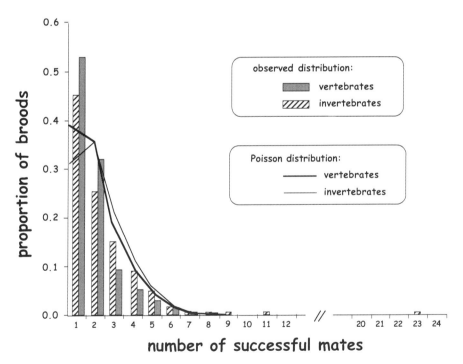

FIGURE 8.9   Empirical frequency distributions of broods in which the gestating invertebrate or vertebrate parent had various numbers of successful mates as deduced from genetic-parentage analyses. Also shown for comparison are the respective Poisson distributions of expected mate counts given the same mean numbers of mates as were documented genetically.

The net effect of such natural-history factors (mating costs and logistical constraints) is to circumscribe successful mate numbers dramatically, even in species with very large broods. Thus, before invoking a selective explanation of genetic polygamy in any focal species, an important question is whether the mean number of successful mates per brood statistically exceeds the rate of mate encounters given the particular biology and ecology of each species. This general kind of sentiment has been expressed previously. For example, after reviewing the literature on multiple paternity in reptiles, Uller and Olsson (2008, 2566) concluded that "The most parsimonious explanation for patterns of multiple paternity is that it represents the combined effect of mate-encounter frequency and conflict over mating rates between males and females driven by large male benefits and relatively small female costs, with only weak selection via indirect benefits."

Thus, for nearly all pregnant and brooding animals (and indeed for essentially all sexual species), an unorthodox but potentially useful null model might envision successful mating opportunities as being in effect relatively "rare and random" events during an individual's lifetime (Avise et al. 2011). Some support for this alternative null perspective on animal mating systems comes from a rather close agreement between empirical histograms of mate numbers per brood and theoretical histograms of mate numbers under statistical Poisson distributions with the same means (fig. 8.9).

Natural Selection Stemming from Pregnancy and Brooding

Chapter 6 described several forms of natural selection that are motivated or amplified by the pregnancy phenomenon in mammals. Some of these expressions of natural selection are likely to be universal for viviparous and brooding species, whereas others may be much more restricted in their taxonomic distributions.

Gestation as Parental Investment

One suite of selective pressures that should apply to all taxa with extended parental care centers on the evolutionary emergence of parent-offspring conflict (Trivers 1974). Whenever a parent invests heavily in pre- or postpartum progeny, selective pressures on the protagonists' genes inevitably swing into motion that in effect can pit parent against child (as well as parent against parent). Haig (2010b) presented a general graphical model (fig. 8.10) of how both conflict and cooperation might be expected between maternal and fetal genes in species with high parental investment (*PI*). Haig (2010b) followed Trivers (1974) in defining *PI* as any investment by the parent in an individual offspring that increases the offspring's chances of surviving (and hence

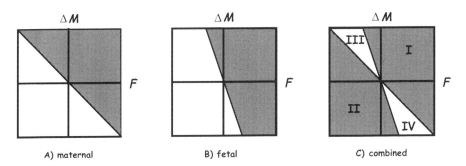

FIGURE 8.10   Haig's (2010b) graphical model of cooperation and conflict between maternal and fetal genes affecting parental investment (*PI*). *F* is a fetus's fitness, and ΔM is the change in a mother's residual fitness (her expected number of progeny, not including the focal fetus) that might result from modifications in the expression of a gene underlying *PI*. Panel A, genes expressed in the mother; panel B, genes expressed in the fetus; panel C, a combination of panels A and B, showing expected zones of maternal-fetal cooperation (regions I and II) and conflict (regions III and IV). These graphs portray outcomes under the assumption that the two focal offspring (a current and a future fetus) are full-sibs ($r=0.5$), but if $r<0.5$ (as would be true under polyandry), the zones of potential maternal-fetal conflict would be even larger. See the text for further explanation.

reproductive success) at the cost of the parent's ability to invest in other offspring. Because *PI* provides a benefit (*B*) to the offspring at a cost (*C*) to the parent's extended fitness, *PI* implies an evolutionary conflict or trade-off such that, from a parent's perspective, additional *PI* is favored whenever $B > C$; whereas from a progeny's perspective, further *PI* is favored whenever $B > rC$, where *r* is the probability that another of the parent's children carries a copy of a randomly chosen gene in the focal offspring. Because $r \leq 0.5$ in all cases, genes in a fetus are generally under selection to demand more *PI* than genes in the parent are selected to supply (Trivers 1974).

In Haig's model (fig. 8.10), the two axes represent changes that a new genetic variant confers on a focal fetus's fitness (*F*) and on a dam's residual fitness (*M*, her future reproduction), respectively. The zero point where the two axes cross represents the evolutionary status quo (i.e., before any new genetic variant arises). Maternal fitness is increased if $F + \Delta M > 0$ (shaded area in panel A), whereas fetal inclusive fitness is increased if $F + r\Delta M > 0$ (shaded area in panel B). In panel C (which combines panels A and B), the shaded regions I and II are zones where a genetic change is favored and disfavored, respectively, regardless of whether it is expressed in the dam or the fetus. By contrast, the unshaded regions III and IV are zones of potential maternal-fetal conflict. In region III, a genetic change is favored if it is caused by a gene expressed in the mother but not by a gene expressed in the fetus (and vice versa for region IV). Overall, regions I and II in panel C thus represent expected zones of maternal-fetal amicability, whereas quadrants III and IV are zones of potential maternal-fetal antagonism.

Although Haig's (2010a) model was offered in the context of imprinted genes and the "placental bed" in mammals (chapter 6), its predictions should apply to many other expressions of parent-offspring conflict (and cooperation) in any species with high parental-investment tactics such as pregnancy or brooding. Four broader sentiments that Haig (2010a) emphasizes are as follows: (1) the evolution of maternal-fetal relations is complex because a dam and her offspring share only some of their genes; (2) the selfish fitness interests of maternal and fetal genes overlap broadly but incompletely; (3) the evolution of any complex maternal-fetal interaction (such as at the placental boundary) is likely to be underlain by genetic changes with some combination of both cooperative and antagonistic etiologies; and (4) parent-offspring relations during and after a pregnancy provide a dynamic evolutionary arena for natural selection because ineluctable fitness trade-offs and compromises exist between the oft-competing genetic interests of the participants.

Alternative Brooding Modes

Although parent-offspring conflicts over finite parental resources are likely to be universal in pregnant and brooding species, exactly how these disputes mechanistically play out must depend in large part on the mode of gestation displayed by a given species or taxonomic group. In external brooders, for example, a parent and an embryo may have little or no direct physical contact, and this clearly limits opportunities for circulating hormones or other effector molecules to mediate metabolic disputes by the physiological or biochemical mechanisms that often characterize species with internal pregnancy (Crespi and Semeniuk 2004; Schrader and Travis 2008). Even among species that are strictly viviparous, the presence or absence of placenta-like structures and the magnitude of material interchange at the parental-fetal interface are likely to affect not only how intergenerational conflicts transpire during each pregnancy but also how they evolve in a given lineage (Reznick et al. 2002; O'Neill et al. 2007; Schrader and Travis 2009; Panhuis et al. 2011). More generally, exactly where a species falls along the matrotrophy-lecithotrophy continuum of parental care can make big differences for the operation of evolutionary forces associated with pregnancy.

Some of the other mechanistic operations that are likely to vary according to gestational mode include immunological interactions between parents and embryos and the syndrome of genetic imprinting for genes involved in fetal growth. Obviously, immunological issues during a pregnancy apply mostly to internal brooders with refined immune systems, whereas genomic imprinting could in principle find at least some form of expression in any species with pregnancy or brooding. Indeed, imprinted genes might be expected to exert their effects on parent-offspring interactions long after a pregnancy itself. For example, the evolution of lactation, age of weaning, and maternal-infant interactions in primates might in principle be influenced by imprinted genes in much the same way as are maternal-fetal interactions in utero. Indeed, from an analysis of human disorders related to imprinted genes, Haig (2010b, 1731) found evidence that "genes of paternal origin, expressed in infants, have been selected to favor more intense suckling than genes of maternal origin."

On the other hand, imprinting effects on genes that influence embryonic growth are not to be expected in taxa such as broadcast spawners, which entirely lack postzygotic parental care. However, most organisms show at least some degree of postzygotic maternal investment in their offspring. Thus, one perplexing question currently demanding further research is why genomic imprinting seems to be confined mostly to mammals and plants. Is there something peculiar about the biology of these organisms that uniquely predisposes them to imprinting, or is genetic imprinting perhaps much more taxonomically widespread than has been adequately documented to date? Moreover, if imprinting does prove to be more taxonomically widespread, what types of genes are involved, and how do their mechanistic operations relate to alternative gestational modes?

Other Contrarian Selective Pressures

Apart from the parent-offspring, parent-parent, and inter-offspring conflicts routinely promoted by pregnancy (chapters 6 and 7), other types of evolutionary disputes may become amplified in viviparous and brooding species. Consider, for example, the frequent opposition of natural selection and sexual selection. Many phenotypic traits associated with pregnancy register evolutionary compromises between these opposing selective forces. An obvious example involving swordtail fish was mentioned in chapter 7. Recall that pregnancy in these viviparous fishes makes females an especially valuable resource for which males actively compete for reproductive access. The choosy females prefer to mate with males possessing longer tails, which thus are promoted by epigamic sexual selection even while being opposed by natural selection because they make swimming more difficult. Fitness trade-offs from such contrarian selection between survival (natural selection) and mating success (sexual selection) must be both common and expressed at many phenotypic levels in species with pregnancy or brooding. Indeed, pregnancy is fundamentally an evolutionary compromise that promotes the survival of the current offspring at the cost of diminishing a parent's future reproductive prospects.

Conflict and Mating Systems

Another evolutionary expectation is that the potential for pregnancy-related conflict should be modulated by the nature of both the mating system and the gestational system. For example, parent-offspring prenatal conflict is generally expected to be most intense in highly polygamous placental species because viviparity offers longer and more intimate contact between embryos and the pregnant parent and because polyandry (or polygyny in male-pregnant species) promotes genomic conflict due to reduced genetic relatedness among littermates. Thus, any conflict that may have originated early in an evolutionary transition from oviparity to viviparity can later become accentuated, especially in lineages that evolve polygamy and placentation.

Footprints of Selection or Phylogeny?

In nearly all comparative research on pregnancy, it becomes necessary at some point to disentangle phenotypic outcomes due to natural selection from phylogenetic legacies that might be fitness neutral or perhaps even maladaptive for their current bearers. Researchers employ at least two general research approaches: (a) functional analyses of extant phenotypes; and (b) phylogenetic character mapping (PCM) to deduce the evolutionary histories of alternative phenotypes (chapter 1). Examples of the former approach are legion and include modern-day attempts to identify the genes, physiological mechanisms, and functional consequences of a wide variety of gestation-related phenomena such as genomic imprinting, placentation, maternal-fetal interactions, and indeed viviparity itself (e.g., Constancia et al. 2002; Haig 2008; Lynch et al. 2008). Examples of the PCM approach likewise are legion and have included attempts to trace the evolutionary histories of pregnancy-germane phenomena such as the degree of placental development in female-pregnant fishes (Reznick et al. 2002), mating systems and sexual selection in male-pregnant fishes (Wilson et al. 2003), and alternative modes of parental care in various taxa (Goodwin et al. 1998; Mank et al. 2005). In the final analysis, a deeper understanding of pregnancy-related selection will require a thorough integration of adaptive evolutionary reasoning in conjunction with both phylogenetic and ontogenetic dissections of particular phenotypes associated with the gestation of embryos.

SUMMARY

1. Pregnancy in effect amplifies what anisogamy initiated by further truncating female fecundity in viviparous species and thereby making females an even more valuable or limiting reproductive resource for which males often

compete intensely for sexual access. This sexual asymmetry has many consequences for mating decisions and sexual selection on males versus females, and in particular it raises the question of why females (in addition to males) sometimes seek multiple mates. In principle, females might receive either direct (material) or indirect (genetic) benefits from polyandrous matings. Depending on the species, direct benefits might include nuptial gifts, access to better territories, reduced sexual harassment, and more or better assistance in rearing offspring. Possible indirect benefits to female fitness include fertilization insurance, higher genetic diversity within each resulting brood as a bet-hedging tactic in variable environments, better paternal genes for progeny, or any of several other such bonuses from multiple mating.

2. Molecular markers and genetic parentage analyses are ideally suited for appraisals of mating behaviors by members of the pregnant sex. In recent years such analyses have unveiled successful mate numbers per brood in numerous vertebrate species that display internal female pregnancy, internal male pregnancy, or external male pregnancy. Empirical summaries of this literature reveal high incidences of successful polygamy in most pregnant species, but they also show that gender-specific restrictions on brood size imposed by these three alternative modes of pregnancy only weakly predict polygamy rates by individuals that brood the offspring.

3. For invertebrate animals that brood their young, a review of the genetic-parentage literature likewise reveals typically high incidences of multiple mating by the gestating parent. Again, however, the data provide little support for the hypothesis that clutch size per se has been a primary selective factor underlying the evolution of mating behaviors by the brooders. Instead, the most striking finding in the comparisons of invertebrates and vertebrates was the generally low (and similar) number of successful mates per brood almost irrespective of brood size. Thus, an alternative null perspective on animal mating systems might well emphasize polygamy's many logistical constraints as a useful counterbalance to the field's standard focus on possible fitness advantages from multiple mating.

4. Alternative pregnancy modes also affect the operation of natural selection. However, an overarching evolutionary theme is that all forms of gestation entail at least moderate parental investment (PI) in offspring. For any species with PI, conflict and cooperation between parents and their offspring are inevitable hallmarks of the procreative process. An extensive body of evolutionary theory and empirical evidence addresses the topics of antagonism, fitness trade-offs, and contrarian selection pressures on male and female parents and on both their current and future offspring. Such conflicts sometimes play out in unanticipated ways, as illustrated by the peculiar phenomena of genomic imprinting.

5. In nearly all discussions of gestation in a comparative context, it is important to disentangle the roles of selection and phylogenetic constraint (historical legacy) in generating present-day outcomes. Whereas many pregnancy-linked phenotypes have been shaped at least in part by natural and/or sexual selection, others are distributed across extant taxa in ways that also register phylogenetic legacies.

APPENDIX: MOLECULAR-GENETIC PARENTAGE ANALYSIS

Several powerful laboratory techniques and associated computer software programs are available for conducting genetic parentage analyses (i.e., for ascertaining who parented whom in nature) (Avise 2004b). Especially for species with pregnancy or brooding, the logic of such analyses is straightforward. Typically, each embryo within an egg case or a nest or inside a gestating parent is genotyped at each of several highly polymorphic Mendelian loci, as is the custodial adult. Then, by comparing the diploid genotype of each offspring with that of its known or suspected biological parent, at each locus the allele that came from the nonpregnant parent usually becomes evident by subtraction (or default). An example of this approach is diagrammed in the accompanying figure (modified from Avise 2002). In this case, a pregnant female and the 19 embryos she carried were genotyped at a locus that proved to display six different alleles (*A–F*) within the brood. Because the known dam was heterozygous for the *A* and *D* alleles, as expected about 50% of the progeny proved to carry maternal allele *A*, and the other 50% carried maternal allele *D*. Barring de novo mutation, this also means that all other alleles (*B*, *C*, *E*, and *F*) in the brood must be of paternal origin. Thus, at least two different males must have mated with the female and genetically contributed to this litter (assuming that each sire was heterozygous for a distinct pair of alleles not shared by the mother).

By extending this approach to multiple polymorphic loci, the multilocus genotype of the sire(s) can often be deduced, and such genetic data thereby permit estimates of the numbers of fathers and their proportionate contributions to each brood. For male-pregnant species such as pipefishes, the same logic applies except that in this case the biological sire is known, and

APPENDIX TABLE 1 Examples of reported numbers of successful mates and frequencies of multiple mating by the pregnant gender in various animal groups.

Group and Species	No. Broods Examined	Frequency of Multiple Matings	Mean (and Range) No. Mates per Brood	Reference
Bony Fishes				
Internal female pregnancy				
Cymatogaster aggregata, shiner perch	27	96	4.6 (1–8)	Liu & Avise 2011
Embiotoca jacksoni, black surfperch	12	100	3.6 (2–6)	Reisser et al. 2009
Embiotoca lateralis, striped surfperch	12	100	3.5 (2–9)	Reisser et al. 2009
Gambusia holbrooki, mosquito fish	50	86	2.2 (1–3)	Zane et al. 1999
Heterandria formosa, least killifish	36	47	1.5 (NA)	Soucy & Travis 2003
Poecilia reticulata, guppy	22	95	3.0 (1–6)	Hain & Neff 2007
Poecilia reticulata, guppy	101	95	3.5 (1–9)	Neff et al. 2008
Sebastes alutus, Pacific ocean perch	66	71	1.9 (1–4)	Von Doornik et al. 2008
Sebastes inermis, black rockfish	5	20	1.2 (1–2)	Gonzalez et al. 2009
Xiphophorus helleri, green swordtail	69	64	1.8 (1–4)	Tatarenkov et al. 2008
Xiphophorus helleri, green swordtail	14	57	1.7 (1–2)	Simmons et al. 2008
Xiphophorus multilineatus, lined swordtail	18	28	1.4 (1–3)	Luo et al. 2005
Internal male pregnancy				
Hippocampus abdominalis, potbelly seahorse	12	0	1.0 (1)	Wilson & Martin-Smith 2007
Hippocampus angustus, Australian seahorse	15	0	1.0 (1)	Jones, Kvarnemo, et al. 1998

Nerophis ophidion, straight-nosed pipefish	15	0	1.0 (1)	McCoy et al. 2001
Syngnathus auliscus, barred pipefish	7	86	2.1 (1–3)	Wilson 2006
Syngnathus floridae, dusky pipefish	22	73	1.9 (1–3)	Jones & Avise 1997b
Syngnathus floridae, dusky pipefish	52	85	2.3 (1–4)	Mobley & Jones 2007
Syngnathus leptorhynchus, bay pipefish	7	86	2.1 (1–3)	Wilson 2006
Syngnathus scovelli, gulf pipefish	40	3	1.0 (1–2)	Jones & Avise 1997a
Syngnathus typhle, broad-nosed pipefish	30	90	3.1 (1–6)	Jones et al. 1999
Syngnathus typhle, broad-nosed pipefish	38	76	2.7 (1–5)	Rispoli & Wilson 2008
External male pregnancy (nest tending)				
Cottus bairdi, mottled sculpin	23	74	2.8 (1–6)	Fiumera et al. 2002
Etheostoma olmstedi, tessellated darter	13	94	3.2 (1–4)	DeWoody, Fletcher, et al. 2000
Etheostoma virgatum, striped darter	20	100	4.7 (2–7)	Porter et al. 2002
Gobiusculus flavescens, two-spotted goby	21	100	4.3 (2–6)	Mobley et al. 2009
Ictalurus punctatus, channel catfish	5	0	1.0 (1)	Tatarenkov et al. 2006
Lepomis auritus, redbreast sunfish	25	100	3.6 (2–6)	DeWoody, Fletcher, et al. 1998
Lepomis marginatus, dollar sunfish	23	83	2.5 (1–7)	Mackiewicz et al. 2002
Lepomis punctatus, spotted sunfish	30	93	4.4 (1–6)	DeWoody, Fletcher, et al. 2000
Pomatoschistus minutus, sand goby	24	100	3.4 (2–6)	Jones et al. 2001

(continued)

Group and Species	No. Broods Examined	Frequency of Multiple Matings	Mean (and Range) No. Mates per Brood	Reference
Enhydris enhydris, rainbow water snake	4	100	4.3 (3–5)	Voris et al. 2008
Enhydris subtaeniata, Mekong mud snake	2	100	4.0 (3–5)	Voris et al. 2008
Eulamprus heatwolei, southern water skink	17	65	1.7 (1–3)	Morrison et al. 2002
Hydrophis and *Lapennis*, sea snakes	12	0	1.0 (1)	Lukoschek & Avise 2011a, 2011b
Lacerta vivipera, common lizard	58	64	1.7 (1–3)	Laloi et al. 2004
Nerodia sipedon, water snake	45	58	1.8 (1–5)	Prosser et al. 2002
Niveoscinctus microlepidotus, snow skink	8	75	N/A	Olsson et al. 2005
Pseudomoia entrecasteuri, log skink	11	27	1.3 (1–2)	Stapley & Keogh 2006
Salamandra salamandra, fire salamander	21	52	1.7 (1–3)	Steinfartz et al. 2006
Thamnophis elegans, western garter snake	6	50	1.7 (1–3)	Garner & Larsen 2005
Thamnophis sirtalis, garter snake	28	57	1.7 (1–3)	King et al. 2001; Garner et al. 2002 + refs.
Tiliqua rugosa, sleepy lizard	17	24	1.2 (1–2)	Bull et al. 1998
Vipera berus, adder	13	69	2.2 (1–4)	Ursenbacher et al. 2009

Invertebrates

Alpheus angulosus, snapping shrimp	54	22	1.2 (1–2)	Mathews 2007
Ammothea biunguiculata, sea spider	45	69	2.7 (1–9)	Barreto & Avise 2011
Ammothea hilgendorfi, sea spider	13	54	1.8 (1–3)	Barreto & Avise 2008
Anrianta arbustorum, land snail	26	100	3.7 (2–6)	Kupfernagel et al. 2010
Aplysia californica, California sea hare	33	76	2.1 (1–3)	Angeloni et al. 2002
Botryllus schlosseri, ascidian	14	92	3.2 (1–5)	Johnson & Yund 2007
Busycon carica, knobbed whelk	12	92	3.5 (1–7)	Walker et al. 2007
Callichirus islagrande, ghost shrimp	40	20	1.2 (1–3)	Bilodeau et al. 2005
Cancer pagurus, brown crab	18	0	1.0 (1)	McKeown & Shaw 2008
Caridina ensifera, freshwater shrimp	20	100	5.3 (2–11)	Yue & Chang 2010
Chionoecetes opilio, snow crab	7	0	1.0 (1)	Urbani et al. 1998
Ciulfina klassi, praying mantis	7	57	2.1 (1–4)	Umbers et al. 2011
Ciulfina rentzi, praying mantis	8	0	1.0 (1)	Umbers et al. 2011
Crepidula coquimbensis, slipper shell	5	100	4.4 (3–5)	Brante et al. 2011
Crepidula fornicata, slipper shell	6	100	3.3 (2–5)	Cam et al. 2009
Crepidula fornicata, slipper shell	18	78	2.6 (1–5)	Dupont et al. 2006
Crepidula fornicata, slipper shell	12	92	2.8 (1–4)	Proestou et al. 2008
Graneledone boreopacifica, octopus	1	100	2.0 (2)	Voight & Feldheim 2009
Homarus americanus, American lobster	108	13	1.2 (1–3)	Gosselin et al. 2005
Littorina obtusata, flat periwinkle	3	100	5.0 (4–6)	Paterson et al. 2001
Littorina saxatilis, rough periwinkle	4	100	19.3 (15–23)	Panova et al. 2010
Loligo forbesi, veined squid	3	67	1.7 (1–2)	Shaw & Boyle 1997

(continued)

Group and Species	No. Broods Examined	Frequency of Multiple Matings	Mean (and Range) No. Mates per Brood	Reference
Loligo pealeii, inshore squid	5	100	3.0 (2–4)	Buresch et al. 2001
Loligo vulgaris, chokka squid	4	50	2.8 (1–5)	Shaw & Sauer 2004
Munida rugosa, rugose squat lobster	25	84	2.2 (1–3)	Bailie et al. 2011
Munida sarsi, squat lobster	5	100	4.0 (4)	Bailie et al. 2011
Nephrops norvegicus, Norway lobster	11	55	1.1 (1–3)	Streiff et al. 2004
Orconectes placidus, placid crayfish	15	60	1.8 (1–4)	Walker et al. 2002
Petrolisthes cinctipes, porcelain crab	10	80	1.9 (1–3)	Toonen 2004
Procambarus clarkii, red swamp crayfish	30	97	2.6 (1–4)	Yue et al. 2010
Sepioteuthis australis, southern calamari	42	98	2.7 (1–4)	van Camp et al. 2004

All data summarized in this table came from studies that used highly polymorphic molecular markers (typically microsatellite loci) to deduce genetic parentage within the broods carried by pregnant females (or in some cases carried or tended by pregnant males). For a more exhaustive compilation for fishes, see Coleman and Jones 2011.

GLOSSARY

abortion See miscarriage.

adaptation Any phenotypic feature that helps an organism survive and reproduce in a particular environment.

adelphophagy Cannibalism by embryos.

afterbirth Remnants of the placenta expelled by a woman after delivery.

aging See senescence.

allantoic sac See allantois.

allantois A saclike membrane present in the embryos of reptiles, birds, and mammals.

allele Any of the possible forms (or classes of forms) of a gene. A diploid individual carries two alleles at each autosomal gene, and these can either be identical in state (in which case the individual is *homozygous*) or different in state (*heterozygous*). At each autosomal gene, a population of *N* diploid individuals harbors 2*N* alleles, some of whose nucleotide sequences may differ.

allograft An artificial transfer of tissue from one individual to another.

allohormone Any substance that, when transferred directly into the body of a conspecific, induces a physiological response.

alloparental care Care of offspring by individuals other than a parent.

allopatric Inhabiting different geographic areas.

implantation   The attachment of an embryo to the uterine wall.

imprinting, genetic   A situation in which a gene is expressed in progeny when inherited from one parent but not the other.

inbreeding   Mating and reproduction between kin.

inbreeding depression   A loss in genetic fitness due to inbreeding.

incestuous   Pertaining to matings between close kin.

inclusive fitness   An individual's own genetic fitness as well as his or her effects on the genetic fitness of close relatives.

incubation   See gestation, brooding.

infant   A recently born offspring.

infanticide   The killing of a live-birthed child.

infertility   An inability to conceive or reproduce.

inner cell mass   Cells that arise mitotically from a zygote and are precursor to the mammalian embryo.

intercourse   Mating involving intromission.

intersexual selection   See epigamic selection.

intrasexual selection   Competition among members of the same sex for mating access.

intromission   During mating, entry of a male's sexual organ into a female.

in vitro fertilization (IVF)   Syngamy outside the living body.

invertebrate   An animal not possessing a backbone.

isogamy   A condition entailing the fusion of gametes of similar size. See also anisogamy.

iteroparity   Repeated reproductive cycles within an individual's lifetime.

kin   Genetic relatives.

kin selection   A form of natural selection due to individuals favoring the survival and reproduction of genetic relatives.

labor   The process of giving birth in viviparous species.

lactation   The secretion of milk from mammary glands.

lecithotrophy   Embryonic dependence upon yolk for nutrition.

life cycle   The sequence of ontogenetic events from zygote to death; one generation.

linked genes Loci carried on the same chromosome.

litter See clutch, brood.

locus (pl. loci) A gene, a location on a chromosome.

madumnal Used in reference to maternally derived alleles present in offspring.

male The sex that produces relatively small gametes.

marsupium An external brood pouch on the abdomen of most female marsupial mammals.

maternal Pertaining to a female parent.

mating system The pattern by which males and females mate (or their gametes unite) in a population. See also monogamy, polygamy, polyandry, polygyny, polygynandry, promiscuity, outcrossing.

matriline A genetic transmission pathway strictly through females.

matrotrophy Direct embryonic dependence upon the mother for nutritional support.

meiosis The cellular process whereby a diploid cell divides to form haploid gametes.

meiotic Pertaining to meiosis.

menarche The time of a girl's first menstruation.

menopause The cessation of menstruation and the normal termination of fertility as a woman ages.

menses See menstruation.

menstruation The periodic flow of blood and mucosal tissue from the uterus to the outside of a woman's body via the vagina.

metabolism The sum of all physical and chemical processes by which living matter is produced and maintained, and by which cellular energy is made available to an organism.

metamorphosis A marked structural transformation during an individual's development.

microbe A very small organism visible only under a microscope.

microsatellite A locus containing tandem repeats of short nucleotide sequences.

miscarriage The natural expulsion and death of an embryo or fetus; in humans, before approximately the 22nd week of gestation. See also abortion.

mitochondrial DNA (mtDNA) A small and typically circular genome housed in the mitochondrion.

mitochondrion (pl. mitochondria) A cytoplasmic organelle that is the site of key metabolic pathways involved in producing cellular energy.

mitosis A process of cell division that produces daughter cells with the same chromosomal constitution as the parent cell.

mitotic Pertaining to mitosis.

molecular clock An evolutionary timepiece based on the evidence that genes or proteins tend to accumulate mutational differences at roughly constant rates in particular lineages.

molecular marker Any nucleic acid or protein used to deduce the genetic relationships or movements of organisms.

monogamy A mating system in which each male and each female has only one mate.

monogenic Attributable to or pertaining to a single genetic locus.

monophyletic Of single evolutionary origin.

monotocous Referring to pregnancies that produce singleton progeny.

monotreme An egg-laying mammal in the Monotremata.

monozygotic twins Twins that trace back to a single fertilized egg.

morphology The visible structures of organisms.

mouthbrooding Parental care of eggs and embryos in the oral cavity.

multicellular Composed of two or more cells.

mutation A change in the genetic constitution of an organism.

natural history The study of nature including organisms and natural phenomena.

natural selection The differential contribution by individuals of different genotypes to the next generation.

nepotism Favoritism directed toward kin.

niche The ecological role of a species in a natural community; an organism's way of making a living.

norm of reaction The range of phenotypes normally produced by a given genotype.

nucleic acid   See deoxyribonucleic acid and ribonucleic acid.

nucleotide   A unit of DNA or RNA consisting of a nitrogenous base, a pentose sugar, and a phosphate group.

nucleus (pl. nuclei)   A portion of a cell bounded by a membrane and containing chromosomes.

obstetrician   A physician who specializes on healthcare during pregnancy and childbirth.

ontogeny   The course of development and growth of an individual to maturity.

oocyte (unfertilized)   A female gamete, also known as an egg cell, or ovum.

oogenesis   The production of oocytes.

oophagy   The consumption of eggs.

operational sex ratio   The number of males versus females (or their gametes) effectively available for reproduction during the time period under consideration.

organ   A part of an animal, such as the heart, that forms a structural and functional unit.

organelle   A complex, recognizable structure in the cell cytoplasm (such as a mitochondrion or chloroplast).

outcrossing   Mating with another, typically unrelated, individual.

ovary   An ovum producing organ.

oviduct   A small tube leading from the ovary to the uterus.

ovigerous   See gravid.

oviparity   Egg-laying.

oviposition   The passage of an egg from the mother's body to the outside.

ovotestis   An organ that produces both eggs and sperm.

ovoviviparity   A system in which young are delivered alive after having hatched from fertilized eggs inside the mother's body.

ovulation   The release of a mature oocyte from the ovary.

ovule   The structure in seed plants that develops into a seed after fertilization of the egg within it.

ovum   An unfertilized egg.

padumnal   Used in reference to paternally derived alleles present in offspring.

parasite   An organism that at some time in its lifecycle is intimately associated with and harms its host.

parasitoid   A parasite that feeds for part of its lifecycle within a host's body and does not have multiple generations per host generation.

parental care   Tending of offspring by a parent. See also parental investment.

parental investment   Any investment by the parent in an offspring that increases the offspring's chances of surviving (and hence reproductive success) at the cost of the parent's ability to invest in other offspring.

parthenogenesis   The development of an individual from an egg without fertilization.

partum   See parturition.

parturition   The exit of an offspring from its parent's body.

paternal   Pertaining to a male parent.

pathogen   An organism or microorganism that produces a disease.

pedigree   A diagram displaying mating partners and their offspring across generations.

pelagic   Pertaining to the open ocean.

periconception   The period of time encompassing gametogenesis, syngamy, and early zygotic development.

perineotomy   A surgical incision into the perineum.

perineum   The portion of the pelvis occupied by the urogenital passages and rectum.

phallodeum   A phallus-like extension of the cloaca in some amphibians.

phallus   An erect penis.

phenotype   Any morphological, physiological, behavioral, or other such characteristics of an organism.

phenotypic plasticity   Variation among phenotypes not due directly to genetic differences. See also norm of reaction.

pheromone   A chemical message, secreted by an individual, that conveys information to and often elicits a specific response from another individual.

photosynthesis   The biochemical process by which a plant uses light to produce carbohydrates from carbon dioxide and water.

phylogenetic   Pertaining to phylogeny.

phylogenetic character mapping (PCM)   A scientific exercise in which alternative traits are plotted and ancestral states are inferred in a phylogenetic framework.

phylogeny   Evolutionary relationships (historical descent) of a group of organisms or species.

phylogeography   Study of the spatial distributions of genealogical lineages.

piscine   Pertaining to fish.

placenta   A disk-shaped organ on the mother's uterine wall that attaches to the umbilical cord and thereby connects a pregnant dam to her gestating child.

placentation   The process of formation of the placenta.

placentotrophy   The consumption of the placenta.

plankton   Small organisms that float freely in the ocean or other large bodies of water.

planktotrophy   A lifecycle that includes a plankton-feeding stage.

planula   A flat free-swimming larval stage.

pleiotropy   A phenomenon in which a single gene can contribute to more than one phenotype.

poecilogony   Any intraspecific polymorphism in larval lifestyles.

pollen   A male gamete in plants.

pollen competition   Rivalry among pollen grains over fertilization success. See also sperm competition; cryptic female choice.

pollination   The transference of pollen to receptive female parts of a plant.

polyandry   A mating system in which particular females may have multiple mates but each male typically has only one mate. Also, any situation in which a focal female has two or more mates.

polyembryony   The production of genetically identical offspring within a clutch.

polygamy   A mating system in which at least some individuals have multiple mates. See also polyandry, polygyny, polygynandry, promiscuity.

polygenic   Attributable to or pertaining to multiple genetic loci.

polygynandry   A mating system in which both males and females may have several mates each.

polygyny   A mating system in which particular males may have multiple mates but each female typically has only one mate. Also, any situation in which a focal male has two or more mates.

polymorphism   The presence of two or more distinct forms (traits or genotypes) in a population.

polyphyletic   A group of organisms perhaps classified together but tracing to different ancestors.

population   All individuals of a species normally inhabiting a defined area.

postzygotic   Following the formation of a fertilized egg.

preadaptation   See exaptation.

precocious puberty   An early onset of puberty.

preeclampsia   High blood pressure associated with a pregnancy.

preemie   A human baby born prematurely. See also premature birth.

premature birth   Early delivery (in humans, before about 37 weeks of gestation).

preterm birth   See premature birth.

prezygotic   Preceding the formation of a fertilized egg.

procreation   Reproduction.

prokaryote   Any microorganism that lacks a chromosome-containing, membrane-bound nucleus.

prolonged pregnancy   In humans, any gestation that lasts longer than about 41–42 weeks.

promiscuity   An extreme form of polygynandry in which each male and female has many mating partners.

protandry   A type of hermaphroditism in which an individual is first male and then later in life switches to female.

protein   A macromolecule composed of one or more polypeptide chains.

protogyny   A type of hermaphroditism in which an individual is first female and then later in life switches to male.

pseudoplacenta   A structure somewhat analogous to the placenta in mammals.

puberty   The stage of adolescence at which an individual becomes physiologically capable to reproduce.

recombination (genetic)   The formation of new combinations of genes, as for example occurs naturally via meiosis and fertilization.

regulatory gene   A segment of DNA that exerts operational control over the expression of other genes.

ribonucleic acid (RNA)   The genetic material of many viruses, similar in structure to DNA. Also, any of a class of molecules that normally arise in cells from the transcription of DNA.

schizophrenia   Any of a group of psychiatric disorders characterized by withdrawal from reality, illogical thinking, and other intellectual disturbances.

secondary sexual traits   Phenotypic characters other than the primary reproductive organs that are elaborated in one sex and evolved via sexual selection.

semelparity   The occurrence of a single brood during an individual's lifetime.

senescence   A persistent decline with age in the survival probability or reproductive output of an individual due to interior physiological deterioration.

sex allocation   The relative parental investment in male versus female reproductive functions.

sex chromosome   A chromosome in the cell nucleus involved in distinguishing the two genders.

sex ratio   The relative number of males versus females in a population.

sex-role reversal   A situation in which sexual selection operates more strongly on females than on males.

sexual dimorphism   Consistent differences in secondary sexual traits between males and females.

sexual reproduction   Organismal procreation via the generation and fusion of gametes.

sexual selection   Selection pressures arising from intraspecific competition for mates.

sexual selection gradient   See Bateman gradient.

Siamese twins   See conjoined twins.

siblicide   One sibling killing another.

siblings   Offspring within a brood or clutch.

sire   The male parent of particular offspring.

soma   See somatic.

somatic   Of or pertaining to any cell (or body part) in a multicellular organism other than those destined to become gametes.

species (biological)   Groups of actually or potentially interbreeding individuals that are reproductively isolated from other such groups.

sperm   A male gamete in animals.

sperm competition   Competition among sperm for fertilization success.

spermatheca   A female storage organ for sperm.

spermatid   One of the four haploid cells produced during each meiosis in males.

spermatogenesis   The production of sperm.

spermatophore   A packet of sperm.

spermozeugmatum   A spherical ball of sperm, as produced by some fishes.

spontaneous abortion   See miscarriage.

squamate reptiles   Snakes and lizards.

suckle   To nurse on a mother's milk.

superembryonation   See superfetation.

superfetation   A situation in which two or more cohorts of embryos of different developmental stages co-occur in a pregnancy.

sympatric   Inhabiting the same geographic area.

synapomorphy   An evolutionarily derived trait shared by two or more related taxa.

syngamy   The genetic union of a male gamete and a female gamete.

systematics   The comparative study and classification of organisms, particularly with regard to their phylogenetic relationships.

taxon (pl. taxa)   A biotic lineage or entity deemed sufficiently distinct from other such lineages as to be worthy of a formal taxonomic name.

taxonomy   The practice of naming and classifying organisms.

testis   A sperm-producing organ.

tissue   A population of cells of the same type performing the same function.

transcription   The cellular process by which an RNA molecule is formed from a DNA template.

translation   The cellular process by which a polypeptide chain is formed from an RNA template.

trophectoderm   Cells that arise mitotically from a zygote, are not part of the embryo itself, but instead are a precursor to the mammalian placenta.

trophic   Pertaining to mode of feeding.

trophoblast   The nonembryonic part of the blastocyst, later developing into the fetal portion of the placenta.

trophonemata   Villous extensions from the uterine wall in some live-bearing fishes.

trophotaeniae   Special placenta-like structures found in some live-bearing animals.

umbilical cord   A rope-like conduit connecting a pregnant mammal to her gestating child.

ungulate   Any large hoof-bearing, grazing mammal.

urethra   The canal through which urine is discharged.

uterus   A hollow organ in the pelvic region of female mammals wherein the fertilized egg implants and develops during a pregnancy.

vagina   The canal leading from the exterior to the uterus in the reproductive tract of females.

variance   Statistical variation; the mean squared deviation from the mean.

vas deferens   A duct in males that carries sperm from the testis to the urethra.

vasectomy   A contraceptive surgical procedure involving removal of all or part of the vas deferens.

vertebrate   An animal that possesses a backbone.

vitellogenesis   The process of yolk formation in an egg.

viviparity   The production and delivery of live offspring from within the body of a parent.

vivipary   In botany, the precocious development of sexual progeny directly on a parent plant.

W chromosome   In birds, the sex chromosome normally present in females only.

womb   A hollow or space where something is generated. See also uterus.

X chromosome   The sex chromosome normally present as two copies in female mammals (the homogametic sex), but as only one copy in males (the heterogametic sex).

Y chromosome   In mammals, the sex chromosome normally present in males only.

yolk   The part of an egg that serves as the primary source of nourishment for an early embryo.

yolk sac   An extra-embryonic membrane that surrounds the yolk in reptilian and avian eggs.

Z chromosome   The sex chromosome normally present as two copies in male birds (the homogametic sex), but as only one copy in females (the heterogametic sex).

zygote   Fertilized egg; the diploid cell arising from the union of male and female haploid gametes.

REFERENCES CITED

Agarwala B. K., and Bhadra, P. 2010. Comparison of fitness characters of two host plant-based congeneric species of the banana aphid, *Pentalonia nigronervosa* and *P. caladii. J. Insect Sci.* 10:1–13.

Amoroso, E. C. 1960. Viviparity in fishes. *Symp. Zool. Soc. London* 1:153–181.

———. 1968. The evolution of viviparity. *Proc. Roy. Soc. Med.* 61:1188–1200.

Andersson, M. 2005. Evolution of classical polyandry: Three steps to female emancipation. *Ethology* 111:1–23.

Andersson, M., and Iwasa, Y. 1996. Sexual selection. *Trends Ecol. Evol.* 11:53–58.

Andersson, M., and Simmons, L. W. 2006. Sexual selection and mate choice. *Trends Ecol. Evol.* 21:296–302.

Angeloni, L., Bradbury, J. W., and Burton, R. S. 2002. Multiple mating, paternity, and brood size in a simultaneous hermaphrodite, *Aplysia californica. Behav. Ecol.* 14:544–560.

Antczak, D. F. 1989. Maternal antibody responses in pregnancy. *Curr. Opin. Immunol.* 1:1135–1140.

Anzenberger, G. 1992. Monogamous mating systems and paternity in primates. In *Paternity in Primates: Genetic Tests and Theories*, ed. Martin, R. D., Dixson, A. F., and Wickings, E. J., 203–224. Basel: Karger.

Arnold, S. J., and Duvall, D. 1994. Animal mating systems: A synthesis based on selection theory. *Amer. Natur.* 143:317–348.

Arnqvist, G., and Nilsson, T. 2000. The evolution of polyandry: Multiple mating and female fitness in insects. *Anim. Behav.* 60:145–164.

Arnqvist, G., and Rowe, L. 2005. *Sexual Conflict*. Princeton, NJ: Princeton University Press.

Asher, M., Lippmann, T., Epplen, J., et al. 2008. Large males dominate: Ecology, social organization, and mating system of wild cavies, the ancestors of the guinea pig. *Behav. Ecol. Sociobiol.* 62:1509–1521.

Ashworth, C. J., Ross, A. W., and Barrett, P. 1998. The use of DNA fingerprinting to assess monozygotic twinning in Meishan and Landrace x large white pigs. *Reprod. Fertil. Dev.* 10:487–490.

Aubuchon, M., Schulz, L. C., and Schust, D. J. 2011. Preeclampsia: Animal models for a human cure. *Proc. Natl. Acad. Sci. USA* 108:1197–1198.

Avise, J. C. 2000. *Phylogeography: The History and Formation of Species.* Cambridge, MA: Harvard University Press.

———. 2002. *Genetics in the Wild.* Washington, DC: Smithsonian Institution Press.

———. 2004a. *The Hope, Hype, and Reality of Genetic Engineering: Remarkable Stories from Agriculture, Industry, Medicine, and the Environment.* New York: Oxford University Press.

———. 2004b. *Molecular Markers, Natural History, and Evolution*, 2nd ed. Sunderland, MA: Sinauer.

———. 2006. *Evolutionary Pathways in Nature: A Phylogenetic Approach.* New York: Cambridge University Press.

———. 2008. *Clonality: The Genetics, Ecology, and Evolution of Sexual Abstinence in Vertebrate Animals.* New York: Oxford University Press.

———. 2010. *Inside the Human Genome: A Case for Non-Intelligent Design.* New York: Oxford University Press.

———. 2011. *Hermaphroditism: The Biology, Ecology, and Evolution of Dual Sexuality.* New York: Columbia University Press.

Avise, J. C., and Ayala, F. J., eds. 2009. *In the Light of Evolution.* Vol. 3, *Two Centuries of Darwin.* Washington, DC: National Academies Press.

Avise, J. C., Jones, A. G., Walker, D., et al. 2002. Genetic mating systems and reproductive natural histories of fishes: Lessons for ecology and evolution. *Annu. Rev. Genet.* 36:19–45.

Avise, J. C., and Liu, J-X. 2010. Multiple mating and its relationship to alternative modes of gestation in male-pregnant versus female-pregnant fish species. *Proc. Natl. Acad. Sci. USA* 107:18915–18920.

———. 2011. Multiple mating and its relationship to brood size in pregnant fishes versus pregnant mammals and other viviparous vertebrates. *Proc. Natl. Acad. Sci. USA* 108:7091–7095.

Avise, J. C., Power, A. J., and Walker, D. 2004. Genetic sex determination, gender identification, and pseudohermaphroditism in the knobbed whelk, *Busycon carica* (Mollusca Melongenidae). *Proc. Roy. Soc. Lond. B* 271:641–646.

Avise, J. C., Quattro, J. M., and Vrijenhoek, R. C. 1992. Molecular clones within organismal clones: Mitochondrial DNA phylogenies and the evolutionary histories of unisexual vertebrates. *Evol. Biol.* 26:225–246.

Avise, J. C., Tatarenkov, A., and Liu, J-X. 2011. Multiple mating and clutch size in invertebrate brooders versus pregnant vertebrates. *Proc. Natl. Acad. Sci. USA* 108:11512–11517.

Avise, J. C., Trexler, J. C., Travis, J., and Nelson, W. S. 1991. *Poecilia mexicana* is the recent female parent of the unisexual fish *P. formosa. Evolution* 45:1530–1533.

Ayre, D. J., and Miller, K. 2006. Random mating in the brooding coral *Acropora palifera. Marine Ecol. Progr. Ser.* 307:155–160.

Badcock, C., and Crespi, B. 2006. Imbalanced genomic imprinting in brain development: An evolutionary basis for the aetiology of autism. *J. Evol. Biol.* 19:1007–1032.

———. 2008. Battle of the sexes may set the brain. *Nature* 454:1054–1055.

Baer, J. G., and Euzet, L. 1961. Classe des Monogènes. In *Traité de Zoologie*, vol. 4, ed. Grassé, P., 241–325. Paris: Masson.

Bailie, D. A., Hynes, R., and Prodöhl, P. A. 2011. Genetic parentage in the squat lobsters *Munida rugosa* and *M. sarsi* (Crustacea, Anomura, Galatheidae). *Mar. Ecol. Progr. Ser.* 421:173–182.

Bain, B. A., and Govedich, F. R. 2004. Courtship and mating behaviour in the Pycnogonida (Chelicerata: Class Pycnogonida): A summary. *Invert. Reprod. Develop.* 46:63–79.

Bainbridge, R. J. 2000. Evolution of mammalian pregnancy in the presence of the maternal immune system. *Rev. Reprod.* 5:67–74.

Baker, H. G. 1955. Self-compatibility and establishment after "long-distance" dispersal. *Evolution* 9:347–349.

Baker, P. J., Funk, S. M., Bruford, M. W., and Harris, S. 2004. Polygynandry in a red fox population: Implications for the evolution of group living in canids? *Behav. Ecol.* 15:766–778.

Baker, R. J., and Bellis, M. A. 1995. *Human Sperm Competition*. London: Chapman and Hall.

Baker, R. J., Makova, K. D., and Chesser, R. K. 1999. Microsatellites indicate a high frequency of multiple paternity in *Apodemus* (Rodentia). *Mol. Ecol.* 8:107–111.

Balinsky, B. I. 1975. *An Introduction to Embryology*, 4th ed. Philadelphia: Saunders.

Baltz, D. M. 1984. Life history variation among female surfperches (Perciformes: Embiotocidae). *Environ. Biol. Fish.* 10:159–171.

Baras, E., Jacobs, B., and Melard, C. 2001. Effects of water temperature on survival, growth and phenotypic sex of mixed (XX-XY) progenies of Nile tilapia, *Oreochromis niloticus. Aquaculture* 192:187–199.

Barreto, F. S., and Avise, J. C. 2008. Polygynandry and sexual size dimorphism in the sea spider *Ammothea hilgendorfi* (Pycnogonida: Ammotheidae), a marine arthropod with brood-carrying males. *Mol. Ecol.* 17:4149–4160.

———. 2010. Quantitative measures of sexual selection reveal no evidence for sex-role reversal in a sea spider with prolonged paternal care. *Proc. Roy. Soc. Lond. B* 277:2951–2956.

———. 2011. The genetic mating system of a sea spider with male-biased sexual size dimorphism: Evidence for paternity skew despite random mating success. *Behav. Ecol. Sociobiol.* 65:1595–1604.

Basolo, A. L. 1990. Female preference predates the evolution of the sword in swordtail fish. *Science* 250:808–810.

———. 1995. Phylogenetic evidence for the role of pre-existing bias in sexual selection. *Proc. Roy. Soc. Lond. B* 259:307–311.

Basolo, A. L., and Alcaraz, G. 2003. The turn of the sword: Length increases male swimming costs in swordtails. *Proc. Roy. Soc. Lond. B* 270:1631–1636.

Bateman, A. J. 1948. Intra-sexual selection in *Drosophila*. *Heredity* 2:349–368.

Baylis, J. R. 1981. The evolution of parental care in fishes, with reference to Darwin's rule of male sexual selection. *Environ. Biol. Fish.* 6:439–449.

Beasley, J. C., Beatty, W. S., Olson, Z. H., and Rhodes, O. E. 2010. A genetic analysis of the Virginia opossum mating system: Evidence of multiple paternity in a highly fragmented landscape. *J. Hered.* 101:368–373.

Beaudet, A. L., and Jiang, Y. 2002. A rheostat model for a rapid and reversible form of imprinting-dependent evolution. *Am. J. Human Genet.* 70:1389–1397.

Beetle, A. A. 1980. Vivipary, proliferation, and phyllody in grasses *J. Range Mgt.* 33:256–261.

Bellemain, E., Swenson, J. E., and Taberlet, P. 2006. Mating strategies in relation to sexually selected infanticide in a non-social carnivore: The brown bear. *Ethology* 112:238–246.

Beltan, L. 1977. La parturition d'un Actinoptérygien de l'Eotrias du nord-ouest de Madagascar. *C.R. Hebd. Acad. Sci. Paris D* 284:2223–2225.

Benirschke, K., Anderson, J. M., and Brownhill, L. E. 1962. Marrow chimerism in marmosets. *Science* 138:513–515.

Berger, A. J. 1953. Three cases of twin embryos in passerine birds. *Condor* 55:273–274.

Berra, T. M., and Humphrey, J. D. 2002. Gross anatomy and histology of the hook and skin of forehood brooding male nurseryfish, *Kurtus gulliveri*, from northern Australia. *Environ. Biol. Fish.* 65:263–270.

Billingham, R. E., and Neaves, W. B. 2005. Exchange of skin grafts among monozygotic quadruplets in armadillos. *J. Exptl. Zool.* 213:257–260.

Bilodeau, A. L., Felder, D. L., and Neigel, J. E. 2005. Multiple paternity in the thalassinidean ghost shrimp, *Callichirus islagrande* (Crustacea: Decapoda, Callianassidae). *Mar. Biol.* 146:381–385.

Birkhead, T. R. 2010. How stupid not to have thought of that: Post-copulatory sexual selection. *J. Zool.* 281:78–93.

Birkhead, T. R., and Møller, A. P. 1992. *Sperm Competition in Birds*. New York: Academic Press.

———. 1993. Female control of paternity. *Trends Ecol. Evol.* 8:100–104.

———, eds. 1998. *Sperm Competition and Sexual Selection*. London: Academic Press.

Birkhead, T. R., and Pizzari, T. 2002. Postcopulatory sexual selection. *Nature Rev. Genet.* 3:262–273.

Bishop, J. D. D., and Pemberton, A. J. 2006. The third way: Spermcast mating in sessile marine invertebrates. *Integr. Comp. Biol.* 46:398–406.

Blackburn, D. G. 1992. Convergent evolution of viviparity, matrotrophy, and specializations for fetal nutrition in reptiles and other vertebrates. *Amer. Zool.* 32:313–321.

———. 1994. Discrepant use of the term "ovoviviparity" in the herpetological literature. *Herpetol. J.* 8:65–71.

———. 1999a. Are viviparity and egg-guarding evolutionarily labile in squamates? *Herpetologica* 55:556–573.

———. 1999b. Placenta and placenta analogue structure and function in reptiles and amphibians. In *Encyclopedia of Reproduction*, vol. 4, ed. Knobil, T. E., and Neill, J. D., 840–847. London: Academic Press.

———. 2000. Classification of the reproductive patterns of amniotes. *Herpetol. Monogr.* 14:371–377.

———. 2005. Evolutionary origins of viviparity in fishes. In *Viviparous Fishes*, ed. Grier, H. J., and Carmen, M., 287–301. Homestead, FL: New Life.

Blackburn, D. G., and Flemming, A. F. 2012. Invasive implantation and intimate placental associations in a placentotrophic African lizard, *Tachylepis ivensi* (Scincidae). *J. Morphol.* 273:137–159.

Blumer, L. S. 1979. Male parental care in the bony fishes. *Quart. Rev. Biol.* 54:149–161.

———. 1982. A bibliography and categorization of bony fishes exhibiting parental care. *Zool. J. Linn. Soc.* 76:1–22.

Bonduriansky, R. 2009. Reappraising sexual coevolution and the sex roles. *PLoS Biol.* 7:1–3.

Bonnano, V. L., and Schulte-Hostedde, A. I. 2009. Sperm competition and ejaculate investment in red squirrels (*Tamiasciurus hudsonicus*). *Behav. Ecol. Sociobiol.* 63:835–846.

Bonnin, A., Goeden, N., Chen, K., Wilson, M. L., King, J., Shih, J. C., et al. 2011. A transient placental source of serotonin for the fetal forebrain. *Nature* 472: 347–350.

Booth, W., Johnson, D. H., Moore, S., Schal, C., and Vargo, E. L. 2010. Evidence for viable, non-clonal but fatherless boa constrictors. *Biol. Lett.* 7:253–256.

Borkowska, A., Borowski, Z., and Krysiuk, K. 2009. Multiple paternity in free-living root voles (*Microtus oeconomus*). *Behav. Proc.* 825:211–213.

Bosolo, A. L. 1990. Female preference predates the evolution of the sword in swordtail fish. *Science* 250:808–810.

———. 1995. Phylogenetic evidence for the role of pre-existing bias in sexual selection. *Proc. Roy. Soc. Lond. Ser. B* 259:307–311.

Bourc'his, D., and Voinnet, O. 2010. A small-RNA perspective on gametogenesis, fertilization, and early zygotic development. *Science* 330:617–622.

Bouteiller, C., and Perrin, N. 2000. Individual reproductive success and effective population size in the greater white-toothed shrew, *Crocidura russula*. *Proc. Roy. Soc. Lond. B* 267:701–705.

Brante, A., Fernandez, M., and Viard, F. 2011. Microsatellite evidence for sperm storage and multiple paternity in the marine gastropod, *Crepiduda coquimbensis*. *J. Exptl. Mar. Biol. Ecol.* 396:83–88.

Breder, M., and Rosen, D. E. 1966. *Modes of Reproduction in Fishes*. Garden City, NY: Natural History Press.

Bryja, J., Patzenhauerova, H., Albrecht, T., Mosansky, L., Stanko, M., and Topka, P. 2008. Varying levels of female promiscuity in four *Apodemus* mice species. *Behav. Ecol. Sociobiol.* 63:251–260.

Buchanan, G. D. 1957. Variation in litter size of nine-banded armadillos. *J. Mamm.* 38:529.

Bucklin, A., Hedgecock, D., and Hand, C. 1984. Genetic evidence of self-fertilization in the sea anemone *Epiactis prolifera*. *Marine Biol.* 84:175–182.

Bull, C. M., Cooper, S. J. B., and Baghurst, B. C. 1998. Social monogamy and extra-pair fertilization in an Australian lizard, *Tiliqua rugosa*. *Behav. Ecol. Sociobiol.* 44:63–72.

Bull, J. J., and Shine, R. 1979. Iteroparous animals that skip opportunities for reproduction. *Amer. Natur.* 114: 296–303.

Bulmer, M. G. 1970. *The Biology of Twinning in Man*. Oxford: Clarendon.

Buresch, K. M., Hanlon, R. T., Maxwell, M. R., and Ring, S. 2001. Microsatellite DNA markers indicate a high frequency of multiple paternity within individual field-collected egg capsules of the squid *Loligo pealeii*. *Mar. Ecol. Progr. Ser.* 210:161–165.

Burt, A., and Trivers, R. 2006. *Genes in Conflict: The Biology of Selfish Genetic Elements*. Cambridge, MA: Belknap.

Burton, C. 2002. Microsatellite analysis of multiple paternity and male reproductive success in the promiscuous snowshoe hare. *Can. J. Zool.* 80:1948–1956.

Busse, K. 1970. Care of young by male *Rhinaderma darwini*. *Copeia* 1970:395.

Byrne, M. 1996. Viviparity and intragonadal cannibalism in the diminutive sea stars *Patiriella vivipara* and *P. parvivipara* (family Asterinidae). *Marine Biol.* 125:551–567.

Byrne, R. J., and Avise, J. C. 2012 Genetic mating systems in elasmobranchs: A case study of the brown smoothhound shark, *Mustelus henlei*. *Marine Biol.* 159:749–756.

Cam, S. L., Pechenik, J. A., Cagnon, M., and Viard, F. 2009. Fast versus slow larval growth in an invasive marine mollusc: Does paternity matter? *J. Hered.* 100:455–464.

Carling, M. D., Wiseman, P. A., and Byers, J. A. 2003. Microsatellite analysis reveals multiple paternity in a population of wild pronghorn antelopes (*Antilocapra americana*). *J. Mammal.* 84:1237–1243.

Carlon, D. B. 2002. Production and supply of larvae as determinants of zonation in a brooding tropical coral. *J. Exptl. Mar. Biol. Ecol.* 268:33–46.

Carmichael, L. E., Szor, G., Berteaux, D., Giroux, M. A., Cameron, C., and Strobeck, C. 2007. Free love in the far north: Plural breeding and polyandry of arctic foxes (*Alopex lagopus*) on Bylot Island, Nunavut. *Can. J. Zool.* 85:338–343.

Carpenter, P. J., Pope, L. C., Creig, C., Dawson, D. A., Rogers, L. M., Erven, K., et al. 2005. Mating system of the Eurasian badger, *Meles meles*, in a high-density population. *Mol. Ecol.* 14:273–284.

Chapman, D. D., Prodohl, P. A., Gelsleichter, J., Manire, C. A., and Shivji, M. S. 2004. Predominance of genetic monogamy by females in a hammerhead shark, *Sphyrna tiburo*: Implications for shark conservation. *Mol. Ecol.* 13:1965–1974.

Chapman, D. D., Shivja, M. S., Louis, E., Sommer, J., Fletcher, H., and Prodohl, P. A. 2007. Virgin birth in a hammerhead shark. *Biol. Lett.* 3:425–427.

Chapple, D. G., and Keogh, J. S. 2005. Complex mating system and dispersal patterns in a social lizard, *Egernia whitii*. *Mol. Ecol.* 14:1215–1227.

Charnov, E. L. 1982. Parent-offspring conflict over reproductive effort. *Amer. Natur.* 119:737.

Chevolot, M., Ellis, J. R., Rijnsdorp, A. D., Stam, W. T., and Olsen, J. L. 2007. Multiple paternity analysis in the thornback ray *Raja clavata*. *J. Hered.* 98:712–715.

Clutton-Brock, T. H. 1991. *The Evolution of Parental Care.* Princeton, NJ: Princeton University Press.

Clutton-Brock, T. H., and Parker, G. A. 1992. Potential reproductive rates and the operation of sexual selection. *Quart. Rev. Biol.* 67:437–456.

Clutton-Brock, T. H., and Vincent, A. C. J. 1991. Sexual selection and the potential reproductive rates of males and females. *Nature* 351:58–60.

Cnaani, A., Lee, B. Y., Zilberman, N., Ozouf-Costaz, C., Hulata, G., Ron, M., et al. 2007. Genetics of sex determination in tilapine species. *Sexual Develop.* 2:43–54.

Cohas, A., Yoccoz, N. G., Da Silva, A., Goossens, B., and Allaine, D. 2006. Extra-pair paternity in the monogamous alpine marmot (*Marmota marmota*): The roles of social setting and female mate choice. *Behav. Ecol. Sociobiol.* 59:597–605.

Colbourne, J. K., Neff, B. D., Wright, J. M., and Gross, M. R. 1996. DNA fingerprinting of bluegill sunfish (*Lepomis macrochirus*) using $(GT)_n$ microsatellites and its potential for assessment of mating success. *Can. J. Fish. Aquat. Sci.* 53:342–349.

Coleman, S. W., and Jones, A. G. 2011. Patterns of multiple paternity and maternity in fishes. *Biol. J. Linn. Soc.* 103:735–760.

Collin, R. 2004. Phylogenetic effects, the loss of complex characters, and the evolution of development in calyptraeid gastropods. *Evolution* 58:1488–1502.

Compagno, L. J. V. 1977. Phyletic relationships of living sharks and rays. *Amer. Zool.* 17:303–322.

Dorgan, L. T., and Clarke, P. E. 1956. Uterus didelphys with double pregnancy. *Amer. J. Obstet. Gynecol.* 72:663–666.

Duellman, W. E., and Trueb, L. 1986. *Biology of Amphibians.* New York: McGraw-Hill.

Dugdale, H. L., Macdonald, D. W., Pope, L. C., and Burke, T. 2007. Polygynandry, extra-group paternity and multiple-paternity litters in European badger (*Meles meles*) social groups. *Mol. Ecol.* 16:5294–5306.

Dulvy, N. K., and Reynolds, J. D. 1997. Evolutionary transitions among egg-laying, live-bearing and maternal inputs in sharks and rays. *Proc. Roy. Soc. Lond. B* 264:1309–1315.

Dupont, L., Richard, J., Paulet, Y. M., Thouzeau, G., and Viard, F. 2006. Gregariousness and protandry promote reproductive insurance in the invasive gastropod *Crepicula fornicata*: evidence from assignment of larval paternity. *Mol. Ecol.* 15:3009–3021.

Dyrynda, P. E. J., and Ryland, J. S. 1982. Reproductive strategies and life histories in the cheilostome marine bryozoans *Chartella papyracea* and *Bugula flabellata. Marine Biol.* 71:241–256.

East, M. L., Burke, T., Wilhelm, K., Greig, C., and Hofer, H. 2003. Sexual conflicts in spotted hyenas: Male and female mating tactics and their reproductive outcome with respect to age, social status and tenure. *Proc. Roy. Soc. Lond. B* 270:1247–1254.

Eberhard, W. G. 1998. Female roles in sperm competition. In *Sperm Competition and Sexual Selection*, ed. Birkhead, T. R., and Møller, A. P., 91–116. London: Academic Press.

———. 2009. Post-copulatory sexual selection: Darwin's omission and its consequences. *Proc. Natl. Acad. Sci. USA* 106 (Suppl.):10025–10032.

Eberle, M., and Kappeler, P. M. 2004. Selected polyandry: Female choice and inter-sexual conflict in a small nocturnal solitary primate (*Microcebus murinus*). *Behav. Ecol. Sociobiol.* 57:91–100.

Ebert, D. 2003. *Sharks, Rays, and Chimaeras of California.* Berkeley: University of California Press.

Edmonds, D. K., Lindsay, K. S., Miller, J. F., Williamson, E., and Wood, P. J. 1982. Early embryonic mortality in women. *Fertil. Steril.* 38:447–453.

Elgar, M. A., and Crispi, B. J., eds. 1992. *Cannibalism: Ecology and Evolution among Diverse Taxa.* Oxford University Press, New York.

Elmqvist, T., and Cox, P. A. 1996. The evolution of vivipary in flowering plants. *Oikos* 77:3–9.

Emlet, R. B. 1990. World patterns of developmental mode in echinoid echinoderms. In *Advances in Invertebrate Reproduction*, ed. Hoshi, M., and Yamashita, O., 329–335. Amsterdam: Elsevier.

Endler, J. A. 1991. Interactions between predators and prey. In *Behavioural Ecology: An Evolutionary Approach*, 3rd ed., ed. Krebs, J. R., and Davies, N. B., 169–196. Oxford: Blackwell Science.

Endler, J. A., and Basolo, A. L. 1998. Sensory ecology, receiver biases and sexual selection. *Trends Ecol. Evol.* 13:415–420.

Engh, A. L., Funk, S. M., Van Horn, C., Scribner, K. T., Bruford, M. W., Libants, S., et al. 2002. Reproductive skew among males in a female-dominated mammalian society. *Behav. Ecol.* 13:193–200.

Ensminger, M. E. 1980. *Dairy Cattle Science*, 2nd ed. Danville, IL: Interstate.

Escomel, E. 1939. La plus jeune mère du monde. *La Presse médicale* 47(38):744.

Farnsworth, E. 2000. The ecology and physiology of viviparous and recalcitrant seeds. *Annu. Rev. Ecol. Syst.* 31:107–138.

Farnsworth, E., and Farrant, J. M. 1998. Reductions in abscisic acid are linked with viviparous reproduction in mangroves. *Amer. J. Bot.* 85:760–769.

Feil, R., and Berger, F. 2007. Convergent evolution of genomic imprinting in plants and mammals. *Trends Genet.* 23:192–199.

Feldheim, K. A., Gruber, S. H., and Ashley, M. V. 2001. Multiple paternity of a lemon shark litter (Chondrichthyes: Carcharhinidae) *Copeia* 2001:781–786.

———. 2004. Reconstruction of parental microsatellite genotypes reveals female polyandry and philopatry in the lemon shark, *Negaprion brevirostris*. *Evolution* 58:2332–2342.

Fitzgerald, G. J. 1992. Filial cannibalism in fishes: Why do parents eat their offspring? *Trends Ecol. Evol.* 7:7–10.

Fiumera, A. C., Porter, B. A., Grossman, G. D., and Avise, J. C. 2002. Intensive genetic assessment of the mating system and reproductive success in a semi-closed population of the mottled sculpin, *Cottus bairdi. Mol. Ecol.* 11:2367–2377.

Fletcher, D. E., Dakin, E. E., Porter, B. A., and Avise, J. C. 2004. Spawning behavior and genetic parentage in the pirate perch (*Aphredoderus sayanus*), a fish with an enigmatic reproductive morphology. *Copeia* 2004:1–10.

Forbes, L. S. 1997. The evolutionary biology of spontaneous abortion in humans. *Trends Ecol. Evol.* 12:446–450.

———. 2002. Pregnancy sickness and embryo quality. *Trends Ecol. Evol.* 17:115–120.

Forstmeier, W., Martin, K., Bolund, E., Schielzeth, H., and Kempenaers, B. 2011. Female extrapair mating behavior can evolve via indirect selection on males. *Proc. Natl. Acad. Sci. USA* 108:10608–10613.

Fowden, L., Sibley, C., Reik, W., and Constancia, M. 2006. Imprinted genes, placental development and fetal growth. *Hormone Res.* 65:50–58.

Frank, S., and Crespi, B. J. 2011. Pathology from evolutionary conflict, with a theory of X chromosome versus autosome conflict over sexually antagonistic traits. *Proc. Natl. Acad. Sci. USA* 108 (Suppl. 2):10886–10893.

Frick, J. E. 1998. Evidence of matrotrophy in the viviparous holothuroid echinoderm *Synaptula hydriformis. Invert. Biol.* 117:169–179.

Frost, J. M., and Moore, G. E. 2010. The importance of imprinting in the human placenta. *PLoS Genet.* 6:e1001015.

Gross, M. R., and Shine, R. 1981. Parental care and mode of fertilization in ecto-thermic vertebrates. *Evolution* 35:775–793.

Guillette, L. J. 1989. The evolution of vertebrate viviparity: Morphological modi-fications and endocrine control. In *Complex Organismal Functions: Integration and Evolution in Vertebrates*, ed. Wake, D. B., and Roth, G., 219–233. New York: Wiley.

Gwynne, D. T. 1984. Courtship feeding increases female reproductive success in bushcrickets. *Nature* 307:361–363.

———. 1991. Sexual competition among females: What causes courtship-role reversal? *Trends Ecol. Evol.* 6:118–121.

Gwynne, D. T., and Simmons, L. W. 1990. Experimental reversal of courtship roles in an insect. *Nature* 346:172–174.

Haataja, R., Karjalainen, M. K., Luukkonen, A., Teramo, K., Puttonen, H., Ojani-emi, M., et al. 2011. Mapping a new spontaneous preterm birth susceptibility gene, *IGF1R*, using linkage, haplotype sharing, and association analysis. *PLoS Genetics* 7: e1001293.

Hagan, H. R. 1948. A brief analysis of viviparity in insects. *J. New York Entomol. Soc.* 56:63–68.

Haig, D. 1990. Brood reduction and optimal parental investment when off-spring differ in quality. *Amer. Natur.* 136:550–556.

———. 1993. Genetic conflicts in human pregnancy. *Quart. Rev. Biol.* 68:495–532.

———. 1996a. Altercation of generations: Genetic conflicts of pregnancy. *Amer. J. Reprod. Immunol.* 35:226–232.

———. 1996b. Gestational drive and the green-bearded placenta. *Proc. Natl. Acad. Sci. USA* 93:6547–6551.

———. 1996c. Placental hormones, genomic imprinting, and maternal-fetal communication. *J. Evol. Biol.* 9:357–380.

———. 1999a. Genetic conflicts of pregnancy and childhood. In *Evolution in Health and Disease*, ed. Stearns, S. C., 77–90. Oxford: Oxford University Press.

———. 1999b. What is a marmoset? *Amer. J. Primatol.* 49:285–296.

———. 2000. The kinship theory of genomic imprinting. *Annu. Rev. Ecol. Syst.* 31:9–32.

———. 2002. *Genomic Imprinting and Kinship.* New Brunswick, NJ: Rutgers Uni-versity Press.

———. 2003. What good is genomic imprinting: The function of parent-specific gene expression. *Nature Rev. Genet.* 4:1–9.

———. 2004a. Evolutionary conflicts in pregnancy and calcium metabolism: A review. *Placenta* 25:S10-S15.

———. 2004b. Genomic imprinting and kinship: How good is the evidence? *Annu. Rev. Genet.* 38:553–585.

———. 2007. Putting up resistance: Maternal-fetal conflict over the control of uteroplacental blood flow. In *Endothelial Medicine*, ed. Aird, W. C., 135–141. Cambridge: Cambridge University Press.

———. 2008. Placental growth hormone-related proteins and prolactin-related proteins. *Trophoblast Res.* 22:S36-S41.

———. 2010a. Fertile soil or no man's land: Cooperation and conflict in the placental bed. In *Placental Bed Disorders*, ed. Pijnenborg, R., Brosens, I., and Romero, R., 165–173. Cambridge: Cambridge University Press.

———. 2010b. Transfers and transitions: Parent-offspring conflict, genomic imprinting, and the evolution of human life history. *Proc. Natl. Acad. Sci. USA* (Suppl. 1) 107:1731–1735.

Hain, S., and Arnaud, P. M. 1992. Notes on the reproduction of high-Antarctic mollusks from the Weddell Sea. *Polar Biol.* 12:303–312.

Hain, T. J. A., and Neff, B. D. 2007. Multiple paternity and kin recognition mechanisms in a guppy population. *Mol. Ecol.* 16:3938–3946.

Hamilton, W. D. 1964. Genetical evolution of social behavior. *J. Theoret. Biol.* 7:1–52.

Hardy, I. C. W. 1995. Protagonists of polyembryony. Reply to Craig et al. from Hardy. *Trends Ecol. Evol.* 10:372.

Harrison, P. L., and Wallace, C. C. 1991. Reproduction, dispersal and recruitment of scleractinian corals. In *Ecosystems of the World*. Vol. 25, *Coral Reefs*, ed. Dubinsky, Z., 133–207. Amsterdam: Elsevier.

Hart, J. A., Hart, M. W., Byrne, M., and Smith, M. J. 1997. Molecular phylogenetic analysis of life-history evolution in asterinid starfish. *Evolution* 5:1848–1861.

Harvey, P. H., and Pagel, M. D. 1991. *The Comparative Method in Evolutionary Biology*. Oxford: Oxford University Press.

Hausfater, G. F., and Hrdy, S. B. 1984. *Infanticide: Comparative and Evolutionary Perspectives*. New York: Aldine.

Hayssen, V. D. 1984. Mammalian reproduction constraints on the evolution of infanticide. In *Infanticide: Comparative and Evolutionary Perspectives*, ed. Hausfater, G. F., and Hrdy, S. B., 105–124. New York: Wenner Gren Foundation.

Hedmark, E., Persson, J., Segerström, P., Landa, A., and Ellegren, H. 2007. Paternity and mating system in wolverines *Gulo gulo*. *Wildlife Biol.* 13 (Suppl. 2):13–30.

Heemstra, P. C., and Greenwood, P. H. 1992. New observations on the visceral anatomy of the late-term fetuses of the living coelacanth fish and the oophagy controversy. *Proc. Roy. Soc. Lond. B* 249:49–55.

Heilmayer, O., Thatje, S., McClelland, C., Conlan, K., and Brey, T. 2008. Changes in biomass and elemental composition during early ontogeny of the Antarctic isopod crustacean *Ceratoserolis trilobitoides*. *Polar Biol.* 31:1325–1331.

Helfman, G. S., Collette, B. B., and Facey, D. E. 1997. *The Diversity of Fishes*. Malden, MA: Blackwell.

Helmuth, B., Veit, R. R., and Holberton, R. 1994. Long-distance dispersal of a subantarctic brooding bivalve (*Gaimardia trapesina*) by kelp-rafting. *Marine Biol.* 120:421–426.

Henson, S. A., and Warner, R. R. 1997. Male and female alternative reproductive behaviors in fishes: A new approach using intersexual dynamics. *Annu. Rev. Ecol. Syst.* 28:571–592.

Herald, E. S. 1961. *Living Fishes of the World*. Garden City, NJ: Doubleday.

Herbert, G. S., and Portell, R. W. 2004. First paleontological record of larval brooding in the calyptraeid gastropod genus *Crepidula* Lamarck, 1799. *J. Paleontol.* 78:424–429.

Herzberg, F., and Herzberg, A. 1962. Observations on reproduction in *Helix aspersa*. *Amer. Midl. Natur.* 68:297–306.

Higgins, L. G. 1954. Prolonged pregnancy (partus serotinus). *Lancet* 267:1154–1156.

Higgs, N. D., Reed, A., Hooke, R., Honey, D., Heilmeyer, O., and Thatje, S. 2009. Growth and reproduction in the Antarctic brooding bivalve *Adacnarca nitens* (Philobryidae) from the Ross Sea. *Marine Biol.* 156:1073–1081.

Highsmith, R. C. 1985. Floating and algal rafting as potential dispersal mechanisms in brooding invertebrates. *Marine Ecol. Progr. Ser.* 25:169–179.

Hill, K., Thomas, K., AbouZahr, C., Walker, N., Say, L., Inoue, M., et al. 2007. Estimates of maternal mortality worldwide between 1990 and 2005: An assessment of available data. *Lancet* 370:1311–1319.

Hoagland, K. 1986. Patterns of encapsulation and brooding in the Calyptraeidae (Prosobranchia: Mesogastropoda). *Amer. Malacol. Bull.* 4:173–183.

Hodge, S. J., Bell, M. B., and Cant, M. A. 2011. Reproductive competition and the evolution of extreme birth synchrony in a cooperative mammal. *Biol. Lett.* 754–756.

Hogarth, P. J. 1976. *Viviparity*. London: Edward Arnold.

Holland, O. J., and Gleeson, D. M. 2005. Genetic characterization of blastocysts and the identification of an instance of multiple paternity in the stoat (*Mustela erminea*). *Conserv. Genet.* 6:855–858.

Holleley, C. E., Dickman, C. R., Crowther, M. S., and Oldroyd, B. P. 2006. Size breeds success: Multiple paternity, multivariate selection and male semelparity in a small marsupial *Antechinus stuartii*. *Mol. Ecol.* 15:3439–3448.

Hook, E. B. 1978. Dietary cravings and aversions during pregnancy. *Amer. J. Clin. Nutr.* 31:1355–1362.

Hubbell, S. P., and Johnson, L. K. 1987. Environmental variance in lifetime mating success, mate choice, and sexual selection. *Amer. Natur.* 130:91–112.

Hudson, R., and Trillmich, F. 2008. Sibling competition and cooperation in mammals: Challenges, developments and prospects. *Behav. Ecol. Sociobiol.* 62:299–307.

Hughes, R. N. 1989. *A Functional Biology of Clonal Animals*. London: Chapman and Hall.

Hurst, L. D. 1990. Parasite diversity and the evolution of diploidy, multicellularity and anisogamy. *J. Theoret. Biol.* 144:429–433.

Hurst, L. D., and Hamilton, W. D. 1992. Cytoplasmic fusion and the nature of sexes. *Proc. Roy. Soc. Lond. B* 247:189–194.

Iavazzo, C. R. 2008. Conception, complicated pregnancy, and labour of gods and heroes in Greek mythology. *Reproductive Biomedicine Online* 17 (Suppl. 1): 11–14.

Ishibashi, Y., and Saitoh, T. 2008. Effect of local density of males on the occurrence of multimale mating in gray-sided voles (*Myodes rufocanus*). *J. Mammal.* 89:388–397.

Jacobs, K. 1971. *Livebearing Aquarium Fishes.* New York: Macmillan.

Jainudeen, M. R., and Hafez, E. S. E. 1980. Reproductive failure in females. In *Reproduction in Farm Animals*, ed. Hafez, E. S. E., 449–470. Philadelphia: Lea and Febiger.

Jennions, M. D., and Petrie, M. 2000. Why do females mate multiply? A review of the genetic benefits. *Biol. Rev.* 75:21–64.

Johannesson, K. 1988. The paradox of Rockall: Why is a brooding gastropod (*Littorina saxatilis*) more widespread than one having a planktonic larval dispersal stage (*L. littorea*)? *Marine Biol.* 99:507–513.

Johannesson, K., Saltin, S. H., Duranovic, I., Havenhand, J. N., and Jonsson, P. R. 2010. Indiscriminate males: Mating behaviour of a marine snail compromised by a sexual conflict? *PloS One* (58):1–7.

Johnson, S. L., and Yund, P. O. 2007. Variation in multiple paternity in natural populations of a free-spawning marine invertebrate. *Mol. Ecol.* 16:3253–3262.

Johnstone, R. W. 1914. Contributions to the study of the early human ovum based upon the investigation of I. A very early ovum embedded in the uterus and II. A very early ovum embedded in the infundibulum of the tube. *J. Obstet. Gynaecol. Br. Emp.* 25:231–276.

Jones, A. G., and Ardren, W. R. 2003. Methods of parentage analysis in natural populations. *Mol. Ecol.* 12:2511–2523.

Jones, A. G., Arguello, J. R., and Arnold, S. J. 2002. Validation of Bateman's principles: A genetic study of sexual selection and mating patterns in the rough-skinned newt. *Proc. Roy. Soc. Lond. B* 269:2533–2539.

Jones, A. G., and Avise, J. C. 1997a. Microsatellite analysis of maternity and the mating system in the Gulf pipefish *Syngnathus scovelli*, a species with male pregnancy and sex-role reversal. *Mol. Ecol.* 6:203–213.

———. 1997b. Polygynandry in the dusky pipefish *Syngnathus floridae* revealed by microsatellite DNA markers. *Evolution* 51:1611–1622.

———. 2001. Mating systems and sexual selection in male-pregnant pipefishes and seahorses: Insights from microsatellite-based studies of maternity. *J. Hered.* 92:150–158.

Jones, A. G., Kvarnemo, C., Moore, G. I., Simmons, L. W., and Avise, J. C. 1998. Microsatellite evidence for monogamy and sex-biased recombination in the Western Australian seahorse *Hippocampus angustus*. *Mol. Ecol.* 7:1497–1505.

Jones, A. G., Moore, G. I., Kvarnemo, C., Walker, D., and Avise, J. C. 2003. Sympatric speciation as a consequence of male pregnancy in seahorses. *Proc. Natl. Acad. Sci. USA* 100:6598–6603.

Jones, A. G., Östlund-Nilsson, S., and Avise, J. C. 1998. A microsatellite assessment of sneaked fertilizations and egg thievery in the fifteenspine stickleback. *Evolution* 52:848–858.

Jones, A. G., and Ratterman, N. L. 2009. Mate choice and sexual selection: What have we learned since Darwin? *Proc. Natl. Acad. Sci. USA* 106 (Suppl.): 10001–10008.

Jones, A. G., Rosenqvist, G., Berglund, A., Arnold, S. J., and Avise, J. C. 2000. The Bateman gradient and the cause of sexual selection in a sex-role-reversed pipefish. *Proc. Roy. Soc. Lond. B* 267:677–680.

Jones, A. G., Rosenqvist, G., Berglund, A., and Avise, J. C. 1999. The genetic mating system of a sex-role-reversed pipefish (*Syngnathus typhle*): A molecular inquiry. *Behav. Ecol. Sociobiol.* 46:357–365.

———. 2005. The measurement of sexual selection using Bateman's principles: An experimental test in the sex-role-reversed pipefish *Syngnathus typhle. Integr. Comp. Biol.* 45:874–884.

Jones, A. G., Small, C. M., Paczolt, K. A., and Ratterman, N. L. 2010. A practical guide to methods of parentage analysis. *Molec. Ecol. Res.* 10:6–30.

Jones, A. G., Walker, D., Kvarnemo, C., Lindström, K., and Avise, J. C. 2001. How cuckoldry can decrease the opportunity for sexual selection: Data and theory from a genetic parentage analysis of the sand goby, *Pomatoschistus minutus. Proc. Natl. Acad. Sci. USA* 98:9151–9156.

Jones, R. E., Gerrard, A. M., and Roth, J. J. 1973. Estrogen and brood pouch formation in the marsupial frog, *Gastrotheca riobambae. J. Exptl. Zool.* 184: 177–184.

Jones, R. K., and Kooistra, K. 2011. Abortion incidence and access to services in the United States. *Persp. Sexual Reproduct. Health* 43:41–50.

Joung, S.-J., Chen, C.-T., Clark, E., Uchida, S., and Huang, W. Y. P. 1996. The whale shark, *Rhincodon typus*, is a live-bearer: 300 embryos found in one "megamamma" supreme. *Environ. Biol. Fish.* 46:219–223.

Kaempf, J. W., Tomlinson, M., Arduza, C., Anderson, S., Campbell, B., Feguson, L. A., et al. 2006. Medical staff guidelines for periviability pregnancy counseling and medical treatment of extremely premature infants. *Pediatrics* 117:22–29.

Kamel, S. J., Grosberg, R. K., and Marshall, D. J. 2010. Family conflicts in the sea. *Trends Ecol. Evol.* 25:442–449.

Kawahara, R., Miya, M., Mabuchi, K., Lavoue, S., Inoue, J. G., Satoh, T. P., et al. 2008. Interrelationships of the 11 gasterosteiform families (sticklebacks, pipefishes, and their relatives): A new perspective based on whole mitogenome sequences from 75 higher teleosts. *Mol. Phylogen. Evol.* 46:224–236.

Keenleyside, M. H. A. 1991. Parental care. In *Cichlid Fishes: Behaviour, Ecology, and Evolution*, 192–208. London: Chapman and Hall.

Keller, L., and Reeve, H. K. 1995. Why do females mate with multiple males? The sexually selected sperm hypothesis. *Adv. Study Behav.* 24:291–315.

Kennis, J., Sluydts, V., Leirs, H., and van Hooft, W. F. 2008. Polyandry and polygyny in an African rodent species, *Mastomys natalensis*. *Mammalia* 72:150–160.

Keverne, E. B. 2010. A mine of imprinted genes. *Science* 466:823–824.

Khan, K. S., Wojdyla, D., Say, L., Gülmezoglu, A. M., and Van Look, P. F. 2006. WHO analysis of causes of maternal death: A systematic review. *Lancet* 367: 1066–1074.

Kikuchi, K., Kai, W., Hosokawa, A., Mizuno, N., Suetake, H., Asahina, K., and Suzuki, Y. 2007. The sex-determining locus of the tiger pufferfish, *Takifugu rubripes*. *Genetics* 175:2039–2042.

King, P. E. 1973. *Pycnogonids*. London: Hutchinson.

King, R. B., Milstead, W. B., Gibbs, H. L., Prosser, M. R., Burghardt, G. M., and McCracken, G. F. 2001. Application of microsatellite DNA markers to discriminate between maternal and genetic effects on scalation and behavior in multiply-sired garter snake litters. *Can. J. Zool.* 79:121–128.

Knapp, R. A., and Sargent, R. C. 1989. Egg-mimicry as a mating strategy in the fantail darter, *Etheostoma flabellare*: Females prefer males with eggs. *Behav. Ecol. Sociobiol.* 25:321–326.

Knowlton, N., and Greenwell, S. R. 1984. Male sperm competition avoidance mechanisms: The influence of female interests. In *Sperm Competition and the Evolution of Animal Mating Systems*, ed. Smith, R. L., 61–84. New York: Academic Press.

Kobmuller, S., Salzburger, W., and Sturmbauer, C. 2004. Evolutionary relationships in the sand-dwelling cichlid lineage of Lake Tanganyika suggest multiple colonizations of rocky habitats and convergent evolution of biparental mouthbrooding. *J. Mol. Evol.* 58:79–96.

Kohda, M., Tanimura, M., Kikue-Nakamura, M., and Yamagashi, S. 1995. Sperm drinking by female catfishes: A novel mode of insemination. *Environ. Biol. Fish.* 42:1–6.

Köhler, D. F., von Rintelen, T., Meyer, A., and Glaubrecht, M. 2004. Multiple origin of viviparity in southeast Asian gastropods (Cerithioidea: Pachychilidae) and its evolutionary implications. *Evolution* 58:2215–2226.

Kraaijeveld-Smit, F. J. L., Ward, S. J., and Temple-Smith, P. D. 2002. Multiple paternity in a field population of a small carnivorous marsupial, the agile antechinus, *Antechinus agilis*. *Behav. Ecol. Sociobiol.* 52:84–91.

Kraus, W. F., Gonzales, M. J., and Vehrencamp, S. L. 1989. Egg development and an evaluation of some of the costs and benefits of paternal care in the belostomatid, *Abedus indentatus* (Heteroptera: Belostomatidae). *J. Kansas Entomol. Soc.* 62:548–562.

Kühnel, S., Reinhard, S., and Kupfer, A. 2010. Evolutionary reproductive morphology of amphibians: An overview. *Bonn Zool. Bull.* 57:119–126.

Kupfer, A., Muller, H., Antoniazzi, M. M., Jared, C., Greven, H., Nussbaum, R. A., and Wilkinson, M. 2006. Parental investment by skin feeding in a caecilian. *Nature* 440:926–929.

Kupfernagel, S., Rusterholz, H.-P., and Baur, B. 2010. Variation in multiple paternity and sperm utilization patterns in natural populations of a simultaneous hermaphrodite land snail. *Biol. J. Linnean Soc.* 99:350–361.

Kvarnemo, C., and Ahnesjö, I. 1996. The dynamics of operational sex ratios and competition for mates. *Trends Ecol. Evol.* 11:404–412.

Laale, H. W. 1984. Polyembryony in teleostean fishes: Double monstrosities and triplets. *J. Fish Biol.* 24:711–719.

Lage, C. R., Petersen, C. W., Forest, D., Barnes, D., Kornfield, I., and Wray, C. 2008. Evidence of multiple paternity in spiny dogfish (*Squalus acanthias*) broods based on microsatellite analysis. *J Fish Biol.* 73:2068–2074.

Laloi, D., Richard, M., Lecomte, J., Massot, M., and Clobert, J. 2004. Multiple paternity in clutches of common lizard *Lacerta vivipara*: Data from microsatellite markers. *Mol. Ecol.* 13:719–723.

Lane, J. E., Boutin, S., Gunn, M. R., Slate, J., and Coltman, D. W. 2008. Female multiple mating and paternity in free-ranging North American red squirrels. *Anim. Behav.* 75:1927–1937.

Larson, S., Christiansen, J., Griffing, D., Ashe, J., Lowry, D. S., and Andrews, K. 2011. Relatedness and polyandry in sixgill sharks, *Hexanchus griseus*, in an urban estuary. *Conserv. Genet.* 12:679–690.

Lee, M. S., and Shine, R. 1998. Reptilian viviparity and Dollo's law. *Evolution* 52:1441–1450.

Lee, P. L. M., and Hays, G. C. 2004. Polyandry in a marine turtle: Females make the best of a bad job. *Proc. Natl. Acad. Sci. USA* 101:6530–6535.

Lessels, C. M., and Parker, G. 1999. Parent-offspring conflict: The full-sib-half-sib fallacy. *Proc. Roy. Soc. Lond. B* 266:1637–1643.

Li, S. K., and Owings, D. H. 1978. Sexual selection in the three-spined stickleback. II. Nest raiding during the courtship phase. *Behaviour* 64:298–304.

Lin, T. T., You, E. M., and Lin, Y. K. 2009. Social and genetic mating systems of the Asian lesser white-toothed shrew, *Crocidura shantungensis*, in Taiwan. *J. Mammal.* 90:1370–1380.

Lindenfors, L., Dalen, P. L., and Angerbjörn, A. 2003. The monophyletic origin of delayed implantation in carnivores and its implications. *Evolution* 57:1952–1956.

Liu, J.-X., and Avise, J. C. 2011. High degree of multiple paternity in the viviparous shiner perch, *Cymatogaster aggregata*, a fish with long-term female sperm storage. *Mol. Ecol.* 158:893–901.

Locket, N. A. 1980. Some advances in coelacanth biology. *Proc. Roy. Soc. Lond. B* 208:265–307.

Logerwell, E., and Ohman, M. D. 1999. Egg-brooding, body size, and predation risk in planktonic marine copepods. *Oecologia* 121:426–431.

Lombardi, J., and Wourms, J. P. 1979. Structure, function, and evolution of trophotaeniae, placental analogues of viviparous fishes. *Amer. Zool.* 19:976.

Long, J. A. 2011. Dawn of the deed: Fish fossils push back the origin of copulation in backboned animals and suggest that it was a turning point in our evolution. *Sci. Amer.* (January), 34–39.

Long, J. A., Trinajstic, K., Young, G. C., and Senden, T. 2008. Live birth in the Devonian period. *Nature* 453:650–652.

Lorch, P. D., Bussiere, L. F., and Gwynne, D. T. 2008. Quantifying the potential for sexual dimorphism using upper limits on Bateman gradients. *Behaviour* 145:1–24.

Loughry, W. J., Dwyer, G. M., and McDonald, C. M. 1998a. Behavioral interactions between juvenile nine-banded armadillos (*Dasypus novemcinctus*) in staged encounters. *Amer. Midland Natur.* 139:125–132.

Loughry, W. J., Prodöhl, P. A., McDonough, C. M., and Avise, J. C. 1998b. Polyembryony in armadillos. *Amer. Sci.* 86:274–279.

Lukoschek, V., and Avise, J. C. 2011a. Development of ten polymorphic microsatellite loci for the sea snake *Hydrophis elegans* (Elapidae: Hydrophiinae) and cross-species amplification for fifteen more marine hydrophine species. *Conserv. Genet. Res.* 3:497–501.

———. 2011b. Genetic monandry in six viviparous species of true sea snakes. *J. Hered.* 102:347–351.

Lund, R. 1980. Viviparity and intrauterine feeding in a new holocephalan fish from the Lower Carboniferous of Montana. *Science* 209:697–699.

Luo, J., Sanetra, M., Schart, M., and Meyer, A. 2005. Strong reproductive skew among males in the multiply mated swordtail *Xiphophorus multilineatus* (Teleostei). *J. Hered.* 96:346–355.

Lynch, V. J., Tanzer, A., Wang, Y., Leung, F. C., Gellerson, B., Emera, D., and Wagner, G. P. 2008. Adaptive changes in the transcription factor HoxA-11 are essential for the evolution of pregnancy in mammals. *Proc. Natl. Acad. Sci. USA* 105:14928–14933.

Lynch, V. J., and Wagner, G. P. 2010. Did egg-laying boas break Dollo's Law? Phylogenetic evidence for reversal to oviparity in sand boas (*Eryx*: Boidae). *Evolution* 64:207–216.

MacGillivray, I., Campbell, D. M., and Thompson, B., eds. 1988. *Twinning and Twins*. Chichester, UK: Wiley.

Mack, P. D., Priest, N. K., and Promislow, D. E. L. 2003. Female age and sperm competition: Last-male precedence declines as female age increases. *Proc. Roy. Soc. Lond. B* 270:159–165.

Mackie, G. L. 1978. Are sphaeriid clams ovoviviparous or viviparous? *Nautilus* 92:145–147.

Mackiewicz, M., Fletcher, D. E., Wilkins, S. D., DeWoody, J. A., and Avise, J. C. 2002. A genetic assessment of parentage in a natural population of dollar sunfish (*Lepomis marginatus*) based on microsatellite markers. *Mol. Ecol.* 11:1877–1883.

Mackiewicz, M., Porter, B. A., Dakin, E. E., and Avise, J. C. 2005. Cuckoldry rates in the Molly Miller (*Scartella cristata*, Blenniidae), a hole-nesting marine fish with alternative reproductive tactics. *Mar. Biol.* 148:213–221.

Macnair, M. R., and Parker, G. 1978. Models of parent-offspring conflict. II. Promiscuity. *Anim. Behav.* 26:111–122.

———. 1979. Models of parent-offspring conflict. III. Intrabrood conflict *Anim. Behav.* 27:1202–1209.

Maguire, D. C. 2003. *Sacred Rights: The Case for Contraception and Abortion in World Religions.* New York: Oxford University Press.

Mair, G. C., Beardmore, J. A., and Skibinski, D. O. F. 1980. Experimental evidence for environmental sex determination in Oreochromis species. In *Proceedings of the Second Asian Fisheries Forum*, ed. Hirano, R., and Hanyu, I., 555–558. Manila, Philippines: Asian Fisheries Society.

Mair, G. C., Scott, A. G., Penman, D. J., Beardmore, J. A., and Skibinski, D. O. F. 1991. Sex determination in the genus Oreochromis. 1. Sex reversal, gynogenesis and triploidy in *Oreochromis niloticus. Theoret. Appl. Genet.* 82:144–152.

Majerus, M. E. N. 2003. *Sex Wars: Genes, Bacteria, and Biased Sex Ratios.* Princeton, NJ: Princeton University Press.

Mank, J. E., and Avise, J. C. 2006a. The evolution of reproductive and genomic diversity in ray-finned fishes: Insights from phylogeny and comparative analysis. *J. Fish Biol.* 69:1–27.

———. 2006b. Supertree analyses of the roles of viviparity and habitat in the evolution of atherinomorph fishes. *J. Evol. Biol.* 19:734–740.

———. 2009. Evolutionary diversity and turn-over of sex determination in teleost fishes. *Sexual Develop.* 3:60–67.

Mank, J. E., Promislow, D. E. L., and Avise, J. C. 2005. Phylogenetic perspectives on the evolution of parental care in ray-finned fishes. *Evolution* 59:1570–1578.

———. 2006. Evolution of alternative sex-determining mechanisms in teleost fishes. *Biol. J. Linnean Soc.* 87:83–93.

Markovic, M., and Trisovic, D. 1979. Monozygotic triplets with discordance for some traits. *Eur. J. Orthodontics* 1:189–192.

Markow, T. A., Beall, S., and Matzkin, L. M. 2008. Egg size, embryonic development time and ovoviviparity in *Drosophila* species. *J. Evol. Biol.* 22:430–434.

Marshall, D. J., and Keough, M. J. 1973. The evolutionary ecology of offspring size in marine invertebrates. *Adv. Mar. Biol.* 53:1–60.

Mathews, L. M. 2007. Evidence for high rates of in-pair paternity in the socially monogamous snapping shrimp *Alpheus angulosus. Aquat. Biol.* 1:55–62.

Matthews, L. H. 1955. The evolution of viviparity in vertebrates. *Mem. Soc. Endocrinol.* 4:129–148.

Maynard Smith, J. 1977. Parental investment: A prospective analysis. *Anim. Behav.* 25:1–9.

McCoy, E. E., Jones, A. G., and Avise, J. C. 2001. The genetic mating system and tests for cuckoldry in a pipefish species in which males fertilize eggs and brood offspring externally. *Mol. Ecol.* 10:1793–1800.

McGowan, R., and Martin, C. C. 1997. DNA methylation and genome imprinting in zebrafish, *Danio rerio*: Some evolutionary ramifications. *Cell Biol.* 75: 499–506.

McKay, R. 2011. Remarkable role for the placenta. *Nature* 472:298–299.

McKaye, K. R., and Kocher, T. 1983. Head ramming behaviour by three paedophagous cichlids in Lake Malawi, Africa. *Anim. Behav.* 31:206–210.

McKeown, N. J., and Shaw, P. W. 2008. Single paternity within brood of the brown crab *Cancer pagurus*: A high fecund species with long-term sperm storage. *Mar. Ecol. Progr. Ser.* 368:209–215.

Medawar, P. B. 1953. Some immunological and endocrinological problems raised by the evolution of viviparity in vertebrates. *Symp. Soc. Exptl. Biol.* 7:320–328.

Meffe, G. K., and Snelson, F. F., Jr., eds. 1989. *Ecology and Evolution of Livebearing Fishes (Poeciliidae)*. Englewood Cliffs, NJ: Prentice Hall.

Meier, R., Kotrba, M., and Ferrar, P. 1999. Ovoviviparity and viviparity in the Diptera. *Biol. Rev. Camb. Philos. Soc.* 74:199–258.

Meyer, A., Morrissey, J. M., and Schartl, M. 1994. Recurrent origin of a sexually selected trait in *Xiphophorus* fishes inferred from a molecular phylogeny. *Nature* 368:539–542.

Miles, D. B., Sinervo, B., and Frankino, W. A. 2000. Reproductive burden, locomotor performance, and the cost of reproduction in free-ranging lizards. *Evolution* 54:1386–1395.

Mobley, K. B., Amundsen, T., Forsgren, E., Svensson, P. A., and Jones, A. G. 2009. Multiple mating and a low incidence of cuckoldry for nest-holding males in the two-spotted goby, *Gobiusculus flavescens*. *BMC Evol. Biol.* 9:6.

Mobley, K. B., and Jones, A. G. 2007. Geographical variation in the mating system of the dusky pipefish (*Syngnathus floridae*). *Mol. Ecol.* 16:2596–2606.

Mock, D. W., and Forbes, L. S. 1992. Parent-offspring conflict: A case of arrested development? *Trends Ecol. Evol.* 7:409–413.

Mock, D. W., and Parker, G. A. 1997. *The Evolution of Sibling Rivalry*. Cambridge: Cambridge University Press.

Møller, A. P., and Birkhead, T. R. 1994. The evolution of plumage brightness in birds is related to extrapair paternity. *Evolution* 48:1089–1100.

Möller, A. P., and Jennions, M. D. 2001. How important are direct fitness benefits to sexual selection? *Naturwissenschaften* 88:401–415.

Moore, T., and Haig, D. 1991. Genomic imprinting in mammalian development: A parental tug-of-war. *Trends Genet.* 7:45–49.

Moran, S., Turner, P. D., and O'Reilly, C. 2009. Multiple paternity in the European hedgehog. *J. Zool.* 278:349–353.

Morrison, S., Keogh, S., and Scott, A. 2002. Molecular determination of paternity in a natural population of the multiply mating polygynous lizard *Eulamprus heatwolei*. *Mol. Ecol.* 11:535–545.

Motocq, M. D. 2004. Reproductive success and effective population size in woodrats (*Neotoma macrotis*). *Mol. Ecol.* 13:1635–1642.

Munshi-South, J. 2007. Extra-pair paternity and the evolution of testis size in a behaviorally monogamous tropical mammal, the large treeshrew (*Tupaia tana*). *Behav. Ecol. Sociobiol.* 62:201–212.

Neese, R. M., and Williams, G. C. 1994. *Why We Get Sick*. New York: Random House.

Neff, B. D. 2001. Genetic paternity analysis and breeding success in bluegill sunfish (*Lepomis macrochirus*). *J. Hered.* 92:111–119.

Neff, B. D., and Gross, M. R. 2001. Dynamic adjustment of parental care in response to perceived paternity. *Proc. Roy. Soc. Lond. B* 268:1559–1565.

Neff, B. D., Pitcher, T. E., and Ramnarine, I. W. 2008. Inter-population variation in multiple paternity and reproductive skew in the guppy. *Mol. Ecol.* 17:2975–2984.

Nelson, J. S. 2006. *Fishes of the World*, 4th ed. New York: Wiley.

Newman, H. H. 1913. The natural history of the nine-banded armadillo of Texas. *Amer. Natur.* 47:513–539.

Nielsen, C. L. R., and Nielsen, C. K. 2007. Multiple paternity and relatedness in southern Illinois raccoons (*Procyon lotor*). *J. Mammal.* 88:441–447.

Niethammer, C. 1977. *Daughters of the Earth: The Lives and Legends of American Indian Women*. New York: Simon and Shuster.

Nohara, M., Nakayama, M., Masamoto, H., Nakazato, K., Sakumoto, K., and Kanazawa, K. 2003. Twin pregnancy in each half of a uterus didelphys with a delivery interval of 66 days. *British J. Obstet. Gynecol.* 110:331–332.

Nowak, M. A., and Highfield, R. 2011. *Supercooperators*. New York: Free Press.

Nowak, M. A., Tarnita, C. E., and Wilson, E. O. 2010. The evolution of eusociality. *Nature* 466:1057–1062.

O'Gara, B. W. 1969. Unique aspects of reproduction in the female pronghorn (*Antilocapra americana* Ord). *Am. J. Anat.* 125:217–232.

Ohman, M. D., and Townsend, A. W. 1998. Egg strings in *Euchirella pseudopulchra* (Aetideidae) and comments on constraints on egg brooding in planktonic marine copepods. *J. Marine Sys.* 15:61–69.

O'Keefe, F. R., and Chiappe, L. M. 2011. Viviparity and K-selected life history in a Mesozoic marine plesiosaur (Reptilia, Sauropterygia). *Science* 333:870–873.

Oliveira, R. F., Taborsky, M., and Brockmann, H. J., eds. 2008. *Alternative Reproductive Tactics: An Integrative Approach*. New York: Cambridge University Press.

Olsen, M. W. 1962. Polyembryony in unfertilized turkey eggs. *J. Hered.* 53:125–129.

Olsson, M., Ujvari, B., Wapstra, E., Madsen, T., Shine, R., and Bensch, S. 2005. Does mate guarding prevent rival mating in snow skink lizards? A test using AFLP. *Herpetologica* 61:389–394.

O'Neill, M. J., Lawton, B. R., Mateos, M., Carone, D. M., Ferreri, G. C., Hrbek, T., et al. 2007. Ancient and continuing Darwinian selection on insulin-like growth factor II in placental fishes. *Proc. Natl. Acad. Sci. USA* 104:12404–12409.

Onorato, D. P., Hellgren, E. C., Van Den Bussche, R. A., and Skiles, J. R. 2004. Paternity and relatedness of American black bears recolonizing a desert montane island. *Can. J. Zool.* 82:1201–1210.

O'Riordan, R. M., Myers, A. A., and Cross, T. F. 1992. Brooding in the intertidal barnacles *Chthamalus stellatus* (Poli) and *Chthalamus montagui* Southward in south-western Ireland. *J. Exptl. Mar. Biol. Ecol.* 164:135–145.

Östlund, S., and Ahnesjö, I. 1998. Female fifteen-spined sticklebacks prefer better fathers. *Anim. Behav.* 56:1177–1183.

Ostrovsky, A., Gordon, D. P., and Lidgard, S. 2009. Independent evolution of matrotrophy in the major classes of Bryozoa: Transitions among reproductive patterns and their ecological background. *Marine Ecol. Progr. Ser.* 378:113–124.

Owusu-Frimpong, M., and Hargreaves, J. A. 2000. Incidence of conjoined twins in tilapia after thermal shock induction of polyploidy. *Aquacult. Res.* 31:421–426.

Packard, G. C., Tracy, C. R., and Roth, J. J. 1977. The physiological ecology of reptilian eggs and embryos, and the evolution of viviparity within the class reptilia. *Biol. Rev.* 52:71–105.

Paczolt, K. A., and Jones, A. G. 2010. Post-copulatory sexual selection and sexual conflict in the evolution of male pregnancy. *Nature* 464:401–404.

Page, E. W. 1939. The relation between hydatid moles, relative ischemia of the gravid uterus and the placental origin of eclampsia. *Am. J. Obstet. Gynecol.* 37:291–293.

Page, L. M., and Bart, H. L., Jr. 1989. Egg mimics in darters (Pisces: Percidae). *Copeia* 1989:514–518.

Page, L. M., and Swofford, D. L. 1984. Morphological correlates of ecological specialization in darters. *Environ. Biol. Fish* 11:139–159.

Panhuis, T. M., Broitman-Maduro, G., Uhrig, J., Maduro, M., and Reznick, D. N. 2011. Analysis of expressed sequence tags from the placenta of the live-bearing fish (*Poeciliopsis* Poeciliidae). *J. Hered.* 102:352–361.

Panova, M., Bostrom, J., Hofving, T., Areskoug, T., Eriksson, A., Mehlig, B., et al. 2010. Extreme female promiscuity in a non-social invertebrate species. *PLoS ONE* 5(3): e9640. doi:10.1371/journal.pone.0009640.

Parker, G. A. 1970. Sperm competition and its evolutionary consequences in the insects. *Biol. Rev.* 45:525–567.

———. 1985. Models of parent-offspring conflict. V. Effects of the behaviour of the two parents. *Anim. Behav.* 33:519–533.

———. 1990. Sperm competition games: Raffles and role. *Proc. Roy. Soc. Lond. B* 242:120–126.

Parker, G. A., Baker, R. R., and Smith, V. G. F. 1972. The origin and evolution of gamete dimorphism and the male-female phenomenon. *J. Theoret. Biol.* 36:529–553.

Parker, G. A., and Macnair, M. R. 1978. Models of parent-offspring conflict. I. Monogamy. *Anim. Behav.* 26:97–110.

Parker, G. A., Royle, N. J., and Hartley, I. R. 2002. Intrafamilial conflict and parental investment: A synthesis. *Phil. Trans. Roy. Soc. Lond. B* 357:295–307.

Parker, G. A., and Simmons, L. W. 1996. Parental investment and the control of sexual selection: Predicting the direction of sexual competition. *Proc. Roy. Soc. Lond. B* 263:315–321.

Parmigiani, S., and Vom Saal, F. S., eds. 1994. *Infanticide and Parental Care.* New York: Taylor and Francis.

Paterson, I. G., Partridge, V., and Buckland-Nicks, J. 2001. Multiple paternity in *Littorina obtusata* (Gastropoda, Littorinidae) revealed by microsatellite analyses. *Biol. Bull.* 200:261–267.

Pattee, O. H., Mattox, W. G., and Seegar, W. S. 1984. Twin embryos in a peregrine falcon egg. *Condor* 86:352–353.

Payne, R. B. 1979. Sexual selection and intersexual differences in variance of breeding success. *Amer. Natur.* 114:447–452.

Pemberton, J. M., Coltman, D. W., Smith, J. A., and Pilkington, J. G. 1999. Molecular analysis of a promiscuous, fluctuating mating system. *Biol. J. Linn. Soc.* 68:289–301.

Pen, I., Uller, T., Feldmeyer, B., Harts, A., While, G. M., and Wapstra, E. 2010. Climate-driven population divergence in sex-determining systems. *Nature* 468:436–438.

Philipp, D. P., and Gross, M. R. 1994. Genetic evidence for cuckoldry in bluegill *Lepomis macrochirus. Mol. Ecol.* 3:563–569.

Pires, M. N., Arendt, J., and Reznick, D. N. 2010. The evolution of placentas and superfetation in the fish genus *Poecilia* (Cyprinodontiformes: Poeciliidae: subgenera *Micropelia* and *Acanthocephalus*). *Biol. J. Linn. Soc.* 99:784–796.

Plachot, M. 1989. Chromosomal analysis of spontaneous abortions after IVF: A European survey. *Human Reprod.* 4:425–429.

Plaut, I. 2002. Does pregnancy affect swimming performance in female mosquitofish, *Gambusia affinis? Funct. Ecol.* 16:290–295.

Pollux, B. J. A., Pires, M. N., Banet, A. I., and Reznick, D. N. 2009. Evolution of placentas in the fish family Poeciliidae: An empirical study of macroevolution. *Annu. Rev. Ecol. Evol. Syst.* 40:271–289.

Porter, B. A., Fiumera, A. C., and Avise, J. C. 2002. Egg mimicry and allopaternal care: Two-mate attracting tactics by which nesting striped darter (*Etheostoma virgatum*) males enhance reproductive success. *Behav. Ecol. Sociobiol.* 51:350–359.

Portnoy, D. S., Piercy, A. N., Musick, J. A., Burgess, G. H., and Graves, J. E. 2007. Genetic polyandry and sexual conflict in the sandbar shark, *Carcharhinus plumbeus*, in the western North Atlantic and Gulf of Mexico. *Mol. Ecol.* 16:187–197.

Poteaux, C., Baubet, E., Kaminski, G., Brandt, S., Dobson, F. S., and Baudoin, C. 2009. Socio-genetic structure and mating system of a wild boar population. *J. Zool.* 278:116–125.

Poulin, E., Palmer, E. T., and Fral, J. P. 2002. Evolutionary versus ecological success in Antarctic benthic invertebrates. *Trends Ecol. Evol.* 187:218–221.

Préault, M., Chastel, O., Cézilly, F., and Faivre, B. 2005. Male bill colour and age are associated with parental abilities and breeding performance in blackbirds. *Behav. Ecol. Sociobiol.* 58:497–505.

Prodöhl, P. A., Loughry, W. J., McDonough, C. M., Nelson, W. S., and Avise, J. C. 1996. Molecular documentation of polyembryony and the micro-spatial dispersion of clonal sibships in the nine-banded armadillo. *Proc. Roy. Soc. Lond. B* 263:1643–1649.

Prodöhl, P. A., Loughry, W. J., McDonough, C. M., Nelson, W. S., Thompson, E. A., and Avise, J. C. 1998. Genetic maternity and paternity in a local population of armadillos assessed by microsatellite DNA markers and field data. *Amer. Natur.* 151:7–19.

Proestou, D. A., Goldsmith, M. R., and Twombly, S. 2008. Patterns of male reproductive success in *Crepidula fornicata* provide new insight for sex allocation and optimal sex change. *Biol. Bull.* 214:194–202.

Profet, M. 1988. The evolution of pregnancy sickness as protection to the embryo against Pleistocene teratogens. *Evol. Theory* 8:177–190.

———. 1992. Pregnancy sickness as adaptation: A deterrent to maternal ingestion of teratogens. In *The Adapted Mind*, ed. Barkow, J., Cosmides, L., and Tooby, J., 327–365. New York: Oxford University Press.

Prosser, M. R., Weatherhead, P. J., Gibbs, H. L., and Brown, G. P. 2002. Genetic analysis of the mating system and opportunity for sexual selection in northern water snakes (*Nerodia sipedon*). *Behav. Ecol.* 13:800–807.

Pryke, S. R., Rollins, L. A., and Griffith, S. C. 2010. Females use multiple mating and genetically loaded sperm competition to target compatible genes. *Science* 329:964–966.

Raff, R. A. 1996. *The Shape of Life: Genes, Development, and the Evolution of Animal Form*. Chicago: University of Chicago Press.

Randall, D. A., Pollinger, J. P., Wayne, R. K., Tallents, L. A., Johnson, P. J., and Macdonald, D. W. 2007. Inbreeding is reduced by female-biased dispersal and mating behavior in Ethiopian wolves. *Behav. Ecol.* 18:579–589.

Raveh, S., Heg, D., Dobson, F. S., Coltman, D. W., Gorrell, J. C., Balmer, A., and Neuhaus, P. 2010. Mating order and reproductive success in male Columbia ground squirrels (*Urocitellus columbianus*). *Behav. Ecol.* 21:537–547.

Redman, C. W. G. 1989. Hypertension in pregnancy. In *Medical Disorders in Obstetric Practice*, 2nd ed., ed. de Swiet, M., 249–305. Oxford, UK: Blackwell.

Reed, C. G. 1991. Bryozoa. *Reproduction of Marine Invertebrates*. VI. *Echnioderms and Lophophorates*, ed. Giese, A. C., and Pearse, J. S., 85–245. Pacific Grove, CA: Boxwood Press.

Reid, D. G. 1990. A cladistic phylogeny of the genus *Littorina* (Gastropoda): Implications for evolution of reproductive strategies and for classification. *Hydrobiologia* 193:1–19.

Reik, W., Constancia, M., Fowden, A., Anderson, N., Dean, W., Ferguson-Smith, A., et al. 2003. Regulation of supply and demand for maternal resources in mammals by imprinted genes. *J. Physiol.* (Lond.) 547:35–44.

Reik, W., and Walter, J. 2001. Genomic imprinting: Parental influence on the genome. *Nature Rev. Genet.* 2:21–32.

Reisser, C. M. O., Beldade, R., and Bernardi, G. 2009. Multiple paternity and competition in sympatric congeneric reef fishes, *Embiotoca jacksoni* and *E. lateralis*. *Mol. Ecol.* 18:1504–1510.

Renfree, M. B., and Shaw, G. 2000. Diapause. *Annu. Rev. Physiol.* 62:353–375.

Reynolds, J. D., Goodwin, N. B., and Freckleton, R. P. 2002. Evolutionary transitions in parental care and live bearing in vertebrates. *Phil. Trans. Roy. Soc. Lond. B* 357:269–281.

Reznick, D. N., Mateos, M., and Springer, M. S. 2002. Independent origins and rapid evolution of the placenta in the fish genus *Poeciliopsis*. *Science* 298:1018–1020.

Reznick, D. N., Meredith, R., and Collette, B. B. 2007. Independent evolution of complex life history adaptations in two families of fishes, live-bearing halfbeaks (Zenarchopteridae, Beloniformes) and Poeciliidae (Cyprinodontiformes). *Evolution* 61:2570–2583.

Reznick, D. N., and Miles, D. B. 1989. A review of life history patterns in poeciliid fishes. In *Ecology and Evolution of Livebearing Fishes Poeciliidae*, ed. Meffe, G. K., and Snelson, F. F., Jr., 125–148. Englewood Cliffs, NJ: Prentice Hall.

Richmond, R. H., and Hunter, C. L. 1990. Reproduction and recruitment of corals: Comparisons among the Caribbean, the tropical Pacific, and the Red Sea. *Marine Ecol. Progr. Ser.* 60:185–203.

Rico, C., Kuhnlein, U., and Fitzgerrald, G. J. 1992. Male reproductive tactics in the threespine stickleback: An evaluation by DNA fingerprinting. *Mol. Ecol.* 1:79–87.

Ripley, J. L. 2009. Osmoregulatory role of the paternal brood pouch for two *Syngnathus* species. *Comp. Biochem. Physiol. A* 154:98–104.

Ripley, J. L., and Foran, C. M. 2009. Direct evidence for embryonic uptake of paternally derived nutrients in two pipefishes (Syngnathidae: *Syngnathus* spp.). *J. Comp. Physiol. B* 179:325–333.

Rispoli, V. F., and Wilson, A. B. 2008. Sexual size dimorphism predicts the frequency of multiple mating in the sex-role reversed pipefish *Syngnathus typhle*. *J. Evol. Biol.* 21:30–38.

Roberts, C. J., and Lowe, C. R. 1975. Where have all the conceptions gone? *Lancet* 1:498–499.

Robinette, W. L., Hancock, N. V., and Jones, D. A. 1977. *The Oak Creek Mule Deer Herd in Utah*. Salt Lake City: Utah State Division of Wildlife Resources.

Rogers-Lowery, C. L., and Dimock, R. V. 2006. Encapsulation of attached ectoparasitic glochidia larvae of freshwater mussels by epithelial tissue on fins of naïve and resistant host fish. *Biol. Bull.* 210:51–63.

Rosen, D. E. 1962. Egg retention pattern in evolution. *Natur. Hist.* 71: 46–53.

Rosen, D. E., and Gordon, M. 1953. Functional anatomy and evolution of male genitalia in poeciliid fishes. *Zoologica* 38:1–47.

Ryan, K. K., and Altmann, J. 2001. Selection for male choice based primarily on mate compatibility in the oldfield mouse, *Peromyscus polionotus rhoadsi*. *Behav. Ecol. Sociobiol.* 50:436–440.

Ryland, J. S. 1970. *Bryozoans*. London: Hutchinson.

Salihu, H. M., Shumpert, M. N., Slay, M., Kirby, R. S., and Alexander, G. R. 2003. Childbearing beyond maternal age 50 and fetal outcomes in the United States. *Obstet. Gynecol.* 102:1006–1014.

Saville, K. J., Lindley, A. M., Maries, E. G., Carrier, J. C., and Pratt, H. L. 2002. Multiple paternity in the nurse shark, *Ginglymostoma cirratum*. *Environ. Biol. Fish.* 63:347–351.

Sawyer, R. T. 1971. The phylogenetic development of brooding behaviour in the Hirundinea. *Hydrobiologia* 37:197–204.

Schartl, M., Wilde, B., Schlupp, I., and Parzefall, J. 1995. Evolutionary origin of a parthenoform, the Amazon molly, *Poecilia formosa*, on the basis of a molecular genealogy. *Evolution* 49:827–835.

Schindler, J. F., and Hamlett, W. C. 1993. Maternal-embryonic relations in viviparous teleosts. *J. Exptl. Zool.* 266:378–393.

Schluter, D., Price, T., Mooers, A., and Ludwig, D. 1997. Likelihood of ancestor states in adaptive evolution. *Evolution* 51:1699–1711.

Schmidt, J. V., Chen, C. C., Sheikh, S. I., Meejkam, M. G., Norman, B. M., and Joung, S. J. 2010. Paternity analysis in a litter of whale shark embryos. *Endangered Sp. Res.* 12:117–124.

Schrader, M., and Travis, J. 2008. Testing the viviparity-driven-conflict hypothesis: Parent-offspring conflict and the evolution of reproductive isolation in a poeciliid fish. *Amer. Natur.* 172:806–817.

———. 2009. Do embryos influence maternal investment? Evaluating maternal-fetal coadaptation and the potential for parent-offspring conflict in a placental fish. *Evolution* 63:2805–2815.

Schultz, E. T. 1993. Sexual size dimorphism at birth in *Micrometrus minimus* (Embiotocidae): A prenatal cost of reproduction. *Copeia* 1993:456–463.

Schulze, S. R., Rice, S. A., Simon, J. L., and Karl, S. A. 2000. Evolution of poecilogony and the biogeography of North American populations of the polychaete *Streblospio*. *Evolution* 54:1247–1259.

Schwartz, M. L., and Dimock, R. Y. 2001. Ultrastructural evidence for nutritional exchange between brooding unionid mussels and their glochidia larvae. *Invert. Biol.* 120:227–236.

Scott, R. J., and Spielman, M. 2006. Genomic imprinting in plants and mammals: How life history constrains convergence. *Cytogen. Genome Res.* 113:53–67.

Scrimshaw, N. S. 1944a. Superfetation in poeciliid fishes. *Copeia* 1944:180–183.

———. 1944b. Embryonic growth in the viviparous poeciliid, *Heterandria formosa. Biol. Bull.* 87:37–51.

Seibel, B., Robison, B. H., and Haddock, S. H. D. 2005. Post-spawning egg care by a squid. *Nature* 438:929.

Sellier, R. 1955. La viviparité chez les insects. *Ann. Biol.* 31:525–545.

Selvin, S. 1980. Probability of non-paternity determined by multiple allele codominant systems. *Amer. J. Hum. Genet.* 15:997–1008.

Sewell, M. A. 1994. Birth, recruitment and juvenile growth in the intraovarian brooding sea cucumber *Leptosynapta clarki. Marine Ecol. Progr. Ser.* 114:149–156.

Shaw, P. W., and Boyle, P. R. 1997. Multiple paternity within the brood of single females of *Loligo forbesi* (Cephalopoda: Loliginidae), demonstrated with microsatellite DNA markers. *Mar. Ecol. Progr. Ser.* 160:279–282.

Shaw, P. W., and Sauer, W. H. H. 2004. Multiple paternity and complex fertilization dynamics in the squid *Loligo vulgaris reynaudii. Mar. Ecol. Progr. Ser.* 270:173–179.

Sherman, C. D. H. 2008. Mating system variation in the hermaphroditic brooding coral, *Seriatophora hystrix. Heredity* 100:296–303.

Sherman, P. W., and Flaxman, S. M. 2001. Protecting ourselves from food. *Amer. Sci.* 89:142–151.

Shurtliff, Q. R., Pearse, D. E., and Rogers, D. S. 2005. Parentage analysis of the canyon mouse (*Peromyscus crinitus*): Evidence for multiple paternity. *J. Mammal.* 86:531–540.

Shuster, S. M. 2009. Sexual selection and mating systems. *Proc. Natl. Acad. Sci. USA* 106 (Suppl.):10009–10016.

Shuster, S. M., and Wade, M. J. 2003. *Mating Systems and Strategies.* Princeton, NJ: Princeton University Press.

Signer, E. N., Anzenberger, G., and Jeffreys, A. J. 2000. Chimaeric and constitutive DNA fingerprints in the common marmoset (*Callithrix jacchus*). *Primates* 41:49–61.

Sikes, R. S., and Ylönen, H. 1998. Considerations of optimal litter size in mammals. *Oikos* 83:452–465.

Simmons, L. W. 2005. The evolution of polyandry: Sperm competition, sperm selection, and offspring viability. *Annu. Rev. Ecol. Evol. Syst.* 36:125–146.

Simmons, L. W., Beveridge, M., and Evans, J. P. 2008. Molecular evidence for multiple paternity in a feral population of green swordtails. *J. Hered.* 99:610–615.

Simpson, M. J. A., Simpson, A. E., Hooley, J., and Zunz, M. 1981. Infant-related influences on birth intervals in rhesus monkeys. *Nature* 290:49–51.

Sims, M. 2009. *In the Womb: Animals.* Washington, DC: National Geographic Books.

Smiley, S., McEuen, F. S., Chaffee, C., and Krishnan, S. 1991. Echinodermata: Holothuroidea. In *Reproduction of Marine Invertebrates*, vol. 6, ed. Giese, C., Pearse, J. S., and Pearse, V. B., 663–750. Pacific Grove, CA: Boxwood.

Smith, C. L., Rand, C., Schaeffer, B., and Atz, J. W. 1975. *Latimeria*, the living coelacanth, is ovoviviparous. *Science* 190:1105–1106.

Smith, R. L. 1974. Life history of *Abedus herberti* in central Arizona (Hemiptera: Belostomatidae). *Psyche* 81:272–283.

———. 1976. Brooding behavior of a male water bug *Belostoma flumineum* (Hemiptera: Belostomatidae). *J. Kansas Entomol. Soc.* 49:333–343.

———. 1979a. Paternity assurance and altered roles in the mating behaviour of a giant water bug, *Abedus herberti* (Heteroptera: Belostomatidae). *Anim. Behav.* 27:716–725.

———. 1979b. Repeated copulation and sperm precedence: Paternity assurance for a male brooding water bug. *Science* 205:1029–1031.

———, ed. 1984. *Sperm Competition and the Evolution of Animal Mating Systems.* New York: Academic Press.

———. 1997. Evolution of parental care in the giant water bugs (Heteroptera: Belostomatidae). In *Social Behavior in Insects and Arachnids*, ed. Choe, J. C., and Crespi, B. J., 116–149. London: Cambridge University Press.

Snyder, B. F., and Gowaty, P. A. 2007. A reappraisal of Bateman's classic study of intrasexual selection. *Evolution* 61:2457–2468.

Solomon, N. G., Keane, B., Knoch, L. R., and Hogan, P. J. 2004. Multiple paternity in socially monogamous prairie voles *Microtus ochrogaster*. *Can. J. Zool.* 82:1667–1671.

Solter, D. 1988. Differential imprinting and expression of maternal and paternal genomes. *Annu. Rev. Genet.* 22:127–146.

Sorin, A. B. 2004. Paternity assignment for white-tailed deer (*Odocoileus virginianus*): Mating across age classes and multiple paternity. *J. Mammal.* 85:356–362.

Soucy, S., and Travis, J. 2003. Multiple paternity and population genetic structure in natural populations of the poeciliid fish, *Heterandria formosa*. *J. Evol. Biol.* 16:1328–1336.

Stamps, J. A., Metcalf, R., and Krishnan, V. V. 1978. Genetic analysis of parent-offspring conflict. *Behav. Ecol. Sociobiol.* 3:369–392.

Stapley, J., and Keogh, J. S. 2006. Experimental and molecular evidence that body size and ventral colour interact to influence male reproductive success in a lizard. *Ethol. Ecol. Evol.* 18:275–288.

Steinfartz, S., Stemshorn, K., Kuesters, D., and Tautz, D. 2006. Patterns of multiple paternity within and between annual reproduction cycles of the fire salamander (*Salamandra salamandra*) under natural conditions. *J. Zool.* 268:1–8.

Toonen, R. J. 2004. Genetic evidence of multiple paternity of broods in the intertidal crab *Petrolisthes cinctipes*. *Marine. Ecol. Progr. Ser.* 270:259–263.

Tregenza, T., and Wedell, N. 2000. Genetic compatibility, mate choice and patterns of parentage. *Mol. Ecol.* 9:1013–1027.

Trexler, J. C., and DeAngelis, D. L. 2003. Resource allocation in offspring provisioning: An evaluation of the conditions favoring the evolution of matrotrophy. *Amer. Natur.* 162:574–585.

Trillmich, F., and Wolf, J. B. W. 2007. Parent-offspring and sibling conflict in Galápagos fur seals and sea lions. *Behav. Ecol. Sociobiol.* 62:363–375.

Trivers, R. L. 1972. Parental investment and sexual selection. In *Sexual Selection and the Descent of Man, 1871–1971*, ed. Campbell, B., 136–179. Chicago: Aldine.

———. 1974. Parent-offspring conflict. *Amer. Zool.* 14:249–264.

Tudge, C. 2000. *The Variety of Life*. New York: Oxford University Press.

Turner, C. L. 1933. Viviparity superimposed upon ovo-viviparity in the Goodeidae, a family of cyprinodont teleost fishes of the Mexican plateau. *J. Morphol.* 55:207–251.

———. 1937. The trophotaeniae of the Goodeidae, a family of viviparous cyprinodont fishes. *J. Morphol.* 61:495–523.

———. 1940a. Pericardial sac, trophotaeniae, and alimentary tract of embryos of goodeid fishes. *J. Morphol.* 67:271–289.

———. 1940b. Pseudoamnion, pseudochorion, and follicular pseudoplacenta in poeciliid fishes. *J. Morphol.* 67:59–89.

———. 1947. Viviparity in teleost fishes. *Sci. Monthly* 65:508–518.

Tycko, B., and Morrison, I. M. 2002. Physiological functions of imprinted genes. *J. Cell Physiol.* 192:245–258.

Tyler, M. J., and Carter, D. B. 1981. Oral birth of the young of the gastric brooding frog *Rheobatrachus silus*. *Anim. Behav.* 29:280–282.

Tyler, M. J., Shearman, D. J., Franco, R., O'Brien, P., Seamark, R. F., and Kelly, R. 1983. Inhibition of gastric acid secretion in the gastric brooding frog, *Rheobatrachus silus*. *Science* 220:609–610.

Tyndale-Biscoe, H., and Renfree, M. 1987. *Reproductive Physiology of Marsupials*. Cambridge: Cambridge University Press.

Uller, T., and Olsson, M. 2008. Multiple paternity in reptiles: Patterns and process. *Mol. Ecol.* 17:2566–2580.

Umbers, K. D. L., Holwell, G. I., Stow, A. J., and Herberstein, M. E. 2011. Molecular evidence for variation in polyandry among praying mantids (Mantodea: *Ciulfina*). *J. Zool.* 284:40–45.

Urbani, N., Sainte-Marie, B., Sevigny, J.-M., Zadworny, D., and Kuhnlein, U. 1998. Sperm competition and paternity assurance during the first breeding period of female snow crab (*Chionoecetes opilio*) (Brachyura: Majidae). *Canad. J. Fish. Aquat. Sci.* 55:1104–1113.

Ursenbacher, S., Erny, C., and Fumagalli, L. 2009. Male reproductive success and multiple paternity in wild, low-density populations of the adder (*Vipera berus*). *J. Hered.* 100:365–370.

Vallowe, H. H. 1953. Some physiological aspects of reproduction in *Xiphophorus maculatus*. *Biol. Bull.* 104:240–249.

van Camp, L. M., Donnellan, S. C., Dyer, A. R., and Fairweather, P. G. 2004. Multiple paternity in field and captive-laid egg strands of *Sepioteuthis australis* (Cephalopoda: Loliginidae). *Mar. Freshw. Res.* 55:819–823.

van Dongen, P. W. J. 2009. Caesarian section: Etymology and early history. *S. Afr. J. Obstet. Gynecol.* 15:62–66.

Van Doornik, D. M., Parker, S. J., Millard, S. R., Berntson, E. A., and Moran, P. 2008. Multiple paternity is prevalent in Pacific ocean perch (*Sebastes alutus*) off the Oregon coast and is correlated with female size and age. *Environ. Biol. Fish.* 83:269–275.

Vance, R. R. 1973. On reproductive strategies in marine benthic invertebrates. *Amer. Natur.* 107:339–352.

Vandeputte, M., Dupont-Nivet, M., Chavanne, H., and Chatain, B. 2007. A polygenic hypothesis for sex determination in the European sea bass—*Dicentrarchus labrax*. *Genetics* 176:1049–1057.

Vanpé, C., Kjellander, P., Gaillard, J. M., Cosson, J. F., Galan, M., and Hewison, A. J. M. 2009. Multiple paternity occurs with low frequency in the territorial roe deer, *Capreolus capreolus*. *Biol. J. Linn. Soc.* 97:128–139.

Varmuza, S., and Mann, M. 1994. Genomic imprinting: Defusing the ovarian time bomb. *Trends Genet.* 10:118–123.

Veith, W. J. 1979. Reproduction in the live-bearing teleost *Clinus superciliosus*. *S. Afr. J. Zool.* 14:208–211.

———. 1980. Viviparity and embryonic adaptations in the teleost *Clinus superciliosus*. *Can. J. Zool.* 58:1–12.

Veríssimo, A., Grubbs, D., McDowell, J., Musick, J., and Portnoy, D. 2011. Frequency of multiple paternity in the spiny dogfish *Squalus acanthias* in the western North Atlantic. *J. Hered.* 102:88–93.

Veron, J. E. N. 2000. *Corals of the World*. Australia: Australian Institute of Marine Sciences.

Vicari, M. R., Artoni, R. F., Moreira-Filgo, O., and Bertollo, L. A. C. 2008. Diversification of a ZZ/ZW sex chromosome system in Characidium fish (Crenuchidae, Characiformes). *Genetica* 134:311–317.

Vincent, A., Ahnesjö, I., Berglund, A., and Rosenqvist, G. 1992. Pipefishes and seahorses: Are they all sex role reversed? *Trends Ecol. Evol.* 7:237–241.

Voight, J. R., and Feldheim, K. A. 2009. Microsatellite inheritance and multiple paternity in the deep sea octopus *Graneledone boreopacifica* (Mullusca: Cephalopoda). *Invert. Biol.* 128:26–30.

Vonhof, M. J., Barber, D., Fenton, M. B., and Strobeck, C. 2006. A tale of two siblings: Multiple paternity in big brown bats (*Eptesicus fuscus*) demonstrated using microsatellite markers. *Mol. Ecol.* 15:241–247.

Voris, H. K., Karns, D. R., Feldheim, K. A., Kechavarzi, B., and Rinehart, M. 2008. Multiple paternity in the Oriental-Australian rear-fanged watersnakes (Homalopsidae). *Herpetol. Conserv. Biol.* 3:88–102.

Vrijenhoek, R. C., and Schultz, R. J. 1974. Evolution of a trihybrid unisexual fish (*Poeciliopsis*, Poeciliidae). *Evolution* 28:306–319.

Wade, M. J., and Arnold, S. J. 1980. The intensity of sexual selection in relation to male sexual behavior, female choice and sperm precedence. *Anim. Behav.* 28:446–461.

Wade, M. J., and Shuster, S. M. 2004. Sexual selection: Harem size and the variance in male reproductive success. *Amer. Natur.* 164:E83-E89.

———. 2010. Bateman (1948): Pioneer in the measurement of sexual selection. *Heredity* 105:507–508.

Wagner, A. P., Creel, S., Frank, L. G., and Kalinowski, S. T. 2007. Patterns of relatedness and parentage in an asocial, polyandrous striped hyena population. *Mol. Ecol.* 16:4356–4369.

Wagner, W. L., Weller, S., and Sakai, A. 2005. Monograph of *Schiedea* (Caryophyllaceae—Alsinoideae). *Syst. Bot. Monogr.* 72:1–169.

Wake, M. H. 1976. The development and replacement of teeth in viviparous caecilians. *J. Morphol.* 148:33–64.

———. 1977a. Fetal maintenance and its evolutionary significance in the amphibia: Gymnophiona. *J. Herp.* 11:379–386.

———. 1977b. The reproductive biology of caecilians. In *The Reproductive Biology of Amphibians*, ed. Taylor, D. H., and Guttman, S. I., 379–386. New York: Plenum.

———. 1993. Evolution of oviductal gestation in amphibians. *J. Exptl. Zool.* 266:394–413.

Walker, D., Porter, B. A., and Avise, J. C. 2002. Genetic parentage assessment in the crayfish *Orconectes placidus*, a high-fecundity invertebrate with extended maternal brood care. *Mol. Ecol.* 11:2115–2122.

Walker, D., Power, A. J., and Avise, J. C. 2005. Sex-linked markers facilitate genetic parentage analyses in knobbed whelk broods. *J. Hered.* 96:108–113.

Walker, D., Power, A. J., Sweeney-Reeves, M., and Avise, J. C. 2007. Multiple paternity and female sperm usage along egg-case strings of the knobbed whelk, *Busycon carica* (Mollusca; Melongenidae). *Marine Biol.* 151:53–61.

Ward, S. 1992. Evidence for broadcast spawning as well as brooding in the scleractinian coral *Poecillopora damicornis*. *Marine Biol.* 112:641–646.

Weisstein, A. E., Feldman, M. W., and Spencer, H. G. 2002. Evolutionary genetic models of the ovarian time bomb hypothesis for the evolution of genomic imprinting. *Genetics* 162:425–439.

Wells, K. D. 2007. *The Ecology and Behaviour of Amphibians*. Chicago: University of Chicago Press.

Wetzel, R. M. 1985. Taxonomy and distribution of armadillos, Dasypodidae. In *The Evolution and Ecology of Armadillos, Sloths, and Vermilinguas*, ed. Montgomery, G. G., 23–46. Washington, DC: Smithsonian Institution Press.

Weygoldt, P. 1980. Complex brood care and reproductive behavior in captive poison-arrow frogs, *Dendrobates pumilio* O. Schmidt. *Behav. Ecol. Sociobiol.* 7:329–332.

Wiebe, J. P. 1968. The reproductive cycle of the viviparous seaperch, *Cymatogaster aggregata* Gibbons. *Can. J. Zool.* 46:1221–1234.

Wilkins, J. F. 2005. Genomic imprinting and methylation: epigenetic canalization and conflict. *Trends Genet.* 21:356–365.

Wilkins, J. F., and Haig, D. 2003. What good is genomic imprinting: The function of parent-specific expression. *Nature Rev. Genet.* 4:1–10.

Wilkinson, M., Kupfer, A., Marques-Porto, R., Jeffkins, H., Antoniazzi, M. M., and Jared, C. 2008. One hundred million years of skin feeding? Extended parental care in a Neotropical caecilian (Amphibia: Gymnophiona). *Biol. Lett.* 4:358–361.

Williams, B., and Cummings, G. 1953. An unusual case of twins. *J. Obstet. Gynaecol. Brit. Emp.* 60:319–321.

Williams, G. C. 1975. *Sex and Evolution*. Princeton, NJ: Princeton University Press.

Wilson, A. B. 2006. Interspecies mating in sympatric species of *Syngnathus* pipefish. *Mol. Ecol.* 15:809–824.

Wilson, A. B., Ahnesjö, I., Vincent, A., and Meyer, A. 2003. The dynamics of male brooding, mating patterns, and sex roles in pipefishes and seahorses family (Syngnathidae). *Evolution* 57:1374–1386.

Wilson, A. B., and Martin-Smith, K. M. 2007. Genetic monogamy despite social promiscuity in the pot-bellied seahorse (*Hippocampus abdominalis*). *Mol. Ecol.* 16:2345–2352.

Wooller, R. D., Richardson, K. C., Garavanta, C. A. M., Saffer, V. M., and Bryant, K. A. 2000. Opportunistic breeding in the polyandrous honey possum, *Tarsipes rostratus*. *Aust. J. Zool.* 48:669–680.

Wourms, J. P. 1977. Reproduction and development of chondrichthyan fishes. *Amer. Zool.* 17:379–410.

———. 1981. Viviparity: The maternal-fetal relationship in fishes. *Amer. Zool.* 21:473–515.

Wourms, J. P., Grove, B. D., and Lombardi, J. 1988. The maternal-fetal relationship in viviparous fishes. In *Fish Physiology*, vol. 2B, ed. Hoar, W. S., and Randall, D. J., 1–134. San Diego: Academic Press.

Wourms, J. P., and Lombardi, J. 1979. Cell ultrastructure and protein absorption in the trophotaenial epithelium, a placental analogue of viviparous fish embryos. *J. Cell Biol.* 83:399a.

Wusterbarth, T. L., King, R. B., Duvall, M. R., Grayburn, W. S., and Burghardt, G. M. 2010. Phylogenetically widespread multiple paternity in new world natricine snakes. *Herpetol. Conserv. Biol.* 5:86–93.

Yamaguchi, N., Sarno, R. J., Johnson, W. E., O'Brien, S. J., and Macdonald, D. W. 2004. Multiple paternity and reproductive tactics of free-ranging American minks, *Mustela vison*. *J. Mammal.* 85:432–439.

Yasui, Y. 1998. The "genetic benefits" of female multiple mating reconsidered. *Trends Ecol. Evol.* 13:246–250.

———. 2001. Female multiple mating as a genetic bet-hedging strategy when mate choice criteria are unreliable. *Ecol. Res.* 16:605–616.

Yue, G. H., and Chang, A. 2010. Molecular evidence for high frequency of multiple paternity in a freshwater shrimp species *Caridina ensifera*. *PLoS ONE* 5(9):e12721. doi:10.1371/journal.pone.0012721.

Yue, G. H., Li, J. L., Wang, C. M., Xia, J. H., Wang, G. L., and Feng, J. B. 2010. High prevalence of multiple paternity in the invasive crayfish species, *Procambarus clarkii*. *Int. J. Biol. Sci.* 6:107–115.

Zane, L., Nelson, W. S., Jones, A. G., and Avise, J. C. 1999. Microsatellite assessment of multiple paternity in natural populations of a live-bearing fish, *Gambusia holbrooki*. *J. Evol. Biol.* 12:61–69.

Zauner, H., Begemann, G., Mari-Beffa, M., and Meyer, A. 2003. Differential regulation of *msx* genes in the development of the gonopodium, an intromittent organ, and of the sword, a sexually selected trait of swordtail fishes (*Xiphophorus*). *Evol. Dev.* 5:466–477.

Zeh, D. W., and Zeh, J. A. 2000. Reproductive modes and speciation: The viviparity-driven conflict hypothesis. *BioEssays* 22:938–946.

Zeh, J. A., and Zeh, D. W. 1996. The evolution of polyandry I: Intragenomic conflict and genetic incompatibility. *Proc. Roy. Soc. Lond. B* 263:1711–1717.

———. 1997. The evolution of polyandry II: Post-copulatory defenses against genetic incompatibility. *Proc. Roy. Soc. Lond. B* 264:69–75.

———. 2001. Reproductive mode and the genetic benefits of polyandry. *Anim. Behav.* 61:1051–1063.

Zinchenko, V. L., and Ivashin, M. V. 1987. Polyembryony and developmental abnormalities in minke whales (Baleanoptera, Acutorostrata) of the southern hemisphere. *Zoology J.* 66:1975–1976.

INDEX

Figures, illustrations, and tables are indicated by italic page numbers. Boxes are indicated by "b" following the page number.